NUMERICAL SEMIGROUPS

Developments in Mathematics

VOLUME 20

Series Editor:
Krishnaswami Alladi, *University of Florida, U.S.A.*

NUMERICAL SEMIGROUPS

By

J.C. ROSALES
University of Granada, Spain

P.A. GARCÍA-SÁNCHEZ
University of Granada, Spain

 Springer

J.C. Rosales
Department of Algebra
Faculty of Sciences
University of Granada
Campus University Fuentenueva
18071 Granada
Spain
jrosales@ugr.es

P.A. García-Sánchez
Department of Algebra
Faculty of Sciences
University of Granada
Campus University Fuentenueva
18071 Granada
Spain
pedro@ugr.es

ISSN 1389-2177
ISBN 978-1-4614-2456-7   e-ISBN 978-1-4419-0160-6
DOI 10.1007/978-1-4419-0160-6
Springer New York Dordrecht Heidelberg London

Mathematics Subject Classification (2000): 20M14, 13H10, 11DXX

© Springer Science+Business Media, LLC 2009
Softcover reprint of the hardcover 1st edition 2009
All rights reserved. This work may not be translated or copied in whole or in part without the written permission of the publisher (Springer Science+Business Media, LLC, 233 Spring Street, New York, NY 10013, USA), except for brief excerpts in connection with reviews or scholarly analysis. Use in connection with any form of information storage and retrieval, electronic adaptation, computer software, or by similar or dissimilar methodology now known or hereafter developed is forbidden.
The use in this publication of trade names, trademarks, service marks, and similar terms, even if they are not identified as such, is not to be taken as an expression of opinion as to whether or not they are subject to proprietary rights.

Printed on acid-free paper

Springer is part of Springer Science+Business Media (www.springer.com)

Para Loly, Patricia y Carlos

Para María y Alba

Contents

Introduction 1

1   Notable elements 5
    Introduction 5
    1   Monoids and monoid homomorphisms 5
    2   Multiplicity and embedding dimension 6
    3   Frobenius number and genus 9
    4   Pseudo-Frobenius numbers 13
    Exercises 16

2   Numerical semigroups with maximal embedding dimension 19
    Introduction 19
    1   Characterizations 20
    2   Arf numerical semigroups 23
    3   Saturated numerical semigroups 28
    Exercises 31

3   Irreducible numerical semigroups 33
    Introduction 33
    1   Symmetric and pseudo-symmetric numerical semigroups ... 33
    2   Irreducible numerical semigroups with arbitrary multiplicity
        and embedding dimension 38
    3   Unitary extensions of a numerical semigroup 44
    4   Decomposition of a numerical semigroup into irreducibles ... 47
    5   Fundamental gaps of a numerical semigroup 51
    Exercises 53

4   Proportionally modular numerical semigroups 57
    Introduction 57
    1   Periodic subadditive functions 57

2   The numerical semigroup associated to an interval
      of rational numbers 59
3   Bézout sequences 61
4   Minimal generators of a proportionally modular numerical
      semigroup.. 65
5   Modular numerical semigroups 69
6   Opened modular numerical semigroups 71
Exercises 75

5   The quotient of a numerical semigroup by a positive integer 77
Introduction 77
1   Notable elements 78
2   One half of an irreducible numerical semigroup 79
3   Numerical semigroups having a Toms decomposition 83
Exercises 88

**6   Families of numerical semigroups closed under finite
     intersections and adjoin of the Frobenius number** 91
Introduction 91
1   The directed graph of the set of numerical semigroups 91
2   Frobenius varieties 93
3   Intersecting Frobenius varieties 98
4   Systems of generators with respect to a Frobenius variety 99
5   The directed graph of a Frobenius variety 100
Exercises 104

7   Presentations of a numerical semigroup 105
Introduction 105
1   Free monoids and presentations............................. 106
2   Minimal presentations of a numerical semigroup 110
3   Computing minimal presentations 113
4   An upper bound for the cardinality of a minimal presentation 117
Exercises 120

8   The gluing of numerical semigroups 123
Introduction 123
1   The concept of gluing 124
2   Complete intersection numerical semigroups.................. 127
3   Gluing of numerical semigroups 129
4   Free numerical semigroups 132
Exercises 135

9   Numerical semigroups with embedding dimension three 137
      Introduction 137
      1   Numerical semigroups with Apéry sets of unique expression 137
      2   Irreducible numerical semigroups with embedding
           dimension three 141
      3   Pseudo-Frobenius numbers and genus of an embedding
           dimension three numerical semigroup 147
      Exercises 154

10   The structure of a numerical semigroup 155
      Introduction 155
      1   Levin's theorem 155
      2   Structure theorem 159
      3    $\mathfrak{N}$-monoids 162
      Exercises 168

Bibliography 171

List of symbols 177

Index 179

Introduction

Let $\mathbb{N}$ be the set of nonnegative integers. A numerical semigroup is a nonempty subset S of $\mathbb{N}$ that is closed under addition, contains the zero element, and whose complement in $\mathbb{N}$ is finite.

If $n_1, \ldots, n_e$ are positive integers with $\gcd\{n_1, \ldots, n_e\} = 1$, then the set $\langle n_1, \ldots, n_e \rangle = \{\lambda_1 n_1 + \cdots + \lambda_e n_e \mid \lambda_1, \ldots, \lambda_e \in \mathbb{N}\}$ is a numerical semigroup. Every numerical semigroup is of this form.

The simplicity of this concept makes it possible to state problems that are easy to understand but whose resolution is far from being trivial. This fact attracted several mathematicians like Frobenius and Sylvester at the end of the 19th century. This is how for instance the Frobenius problem arose, concerned with finding a formula depending on $n_1, \ldots, n_e$ for the largest integer not belonging to $\langle n_1, \ldots, n_e \rangle$ (see [52] for a nice state of the art on this problem).

During the second half of the past century, numerical semigroups came back to the scene mainly due to their applications in algebraic geometry. Valuations of analytically unramified one-dimensional local Noetherian domains are numerical semigroups under certain conditions, and many properties of these rings can be characterized in terms of their associated numerical semigroups. For a field K, the valuation of the ring $K[[t^{n_1}, \ldots, t^{n_e}]]$ is precisely $\langle n_1, \ldots, n_e \rangle$. This link can be used to construct one-dimensional Noetherian local domains with the desired properties, and it is basically responsible for how some invariants in a numerical semigroup have been termed. Such invariants include the multiplicity, embedding dimension, degree of singularity, type and conductor. Some families of numerical semigroups also were considered partly because of this connection: symmetric numerical semigroups, pseudo-symmetric numerical semigroups, numerical semigroups with maximal embedding dimension and with the Arf property, saturated numerical semigroups, and complete intersections, each having their counterpart in ring theory. A good translator for these concepts between both ring and semigroup theory is [5]. It is worth mentioning that these semigroups are important not only for their applications in algebraic geometry, but also because their definitions appear in a very natural way in the scope of numerical semigroups. One of the aims of this volume is to show this.

J.C. Rosales, P.A. García-Sánchez, *Numerical Semigroups*,
Developments in Mathematics 20, DOI 10.1007/978-1-4419-0160-6_1,
© Springer Science+Business Media, LLC 2009

Recently, the study of factorizations on integral domains has moved to the setting of commutative cancellative monoids (this is mainly due to the fact that addition is not needed to study factorizations into irreducibles). Numerical semigroups are cancellative monoids. Problems of factorizations in a monoid are closely related to presentations of the monoid. By taking advantage of the results obtained in the past decades for the computation of minimal presentations of a numerical semigroup, numerical semigroups have become a nice source of examples in factorization theory. This is not the only connection with number theory. Recently, the study of certain Diophantine modular inequalities gave rise to the concept of proportionally modular numerical semigroups, which are related with the Stern-Brocot tree, and whose finite intersections can be realized as the positive cone of certain amenable C^*-algebras.

Finding the set of factorizations of an element in a numerical semigroup can be done with linear integer programming. We will also show another relation with linear integer programming, by proving that the set of numerical semigroups with given multiplicity is in one-to-one connection with the set of integer points in a rational cone.

From a classic point of view, people working in semigroup theory have been mainly concerned with characterizing families of semigroups via the properties they fulfill. In the last chapter of this monograph, we present several characterizations of numerical semigroups as finitely generated commutative monoids with some extra properties.

At the end of each chapter, the reader will find a series of exercises. Some cover concepts not included in the theory presented in the book, but whose relevance has been highly motivated in the literature, and can be solved by using the tools presented in this monograph. Others are simply thought of as a tool to practice and to deepen the definitions given in the chapter. There is also a series of exercises that covers some recent results, and a reference to where they can be found is given. Sometimes these problems are split in smaller parts so that the readers can produce their own proofs.

Some of the proofs presented in this volume can be performed by using commutative algebra tools. Our goal has been to write a self-contained monograph on numerical semigroups that needs no auxiliary background other than basic integer arithmetic. This is mainly why we have not taken advantage of commutative algebra or algebraic geometry.

Acknowledgments

The authors were supported by the project MTM2007-62346 entitled "Semigrupos numéricos." We would like to thank Marco D'Anna, Maria Bras-Amorós, Antonio Jesús Rodríguez Salas and Micah J. Leamer for their helpful comments and suggestions. Special thanks to Manuel Delgado and Paulo Vasco who carefully read the whole manuscript and to Robert Guralnick for his efficiency as an editor. The second author is also grateful to PRODICOYDE and the ELA Calahonda-Carchuna for giving him mail and e-mail access during July and August 2007. The authors would like to thank Elizabeth Loew, Nathan Brothers, and Brian Treadway for their assistance with our publication.

Chapter 1
Notable elements

Introduction

The study of numerical semigroups is equivalent to that of nonnegative integer solutions to a linear nonhomogeneous equation with positive integer coefficients. Thus it is a classic problem that has been widely treated in the literature (see [12, 13, 22, 28, 42, 101, 102]). Following this classic line, two invariants play a role of special relevance in a numerical semigroup. These are the Frobenius number and the genus. Besides, in the literature one finds many manuscripts devoted to the study of analytically unramified one-dimensional local domains via their value semigroups, which turn out to be numerical semigroups (just to mention some of them, see [5, 6, 19, 27, 32, 44, 105, 107]). Playing along this direction other invariants of a numerical semigroup arise: the multiplicity, embedding dimension, degree of singularity, conductor, Apéry sets, pseudo-Frobenius numbers and type. These invariants have their interpretation in this context, and this is the reason why their names may seem bizarre in the scope of monoids. In this sense the monograph [5] serves as an extraordinary dictionary between these apparently two different parts of Mathematics.

1 Monoids and monoid homomorphisms

Numerical semigroups live in the world of monoids. Thus we spend some time here recalling some basic definitions and facts concerning them.

A *semigroup* is a pair $(S, +)$ with S a set and $+$ a binary operation on S that is associative. All semigroups considered in this book are commutative ($a + b = b + a$ for all $a, b \in S$). For this reason we will not keep repeating the adjective commutative in what follows. Usually we will also omit the binary operation $+$ while referring to a commutative semigroup and will write S instead of $(S, +)$. A subsemigroup T of a semigroup S is a subset that is closed under the binary operation considered on

J.C. Rosales, P.A. García-Sánchez, *Numerical Semigroups,*
Developments in Mathematics 20, DOI 10.1007/978-1-4419-0160-6_2,
© Springer Science+Business Media, LLC 2009

S. Clearly, the intersection of subsemigroups of a semigroup *S* is again a subsemigroup of *S*. Thus given *A* a nonempty subset of *S*, the smallest subsemigroup of *S* containing *A* is the intersection of all subsemigroups of *S* containing *A*. We denote this semigroup by $\langle A \rangle$, and call it the *subsemigroup generated* by *A*. It follows easily that

$$\langle A \rangle = \{ \lambda_1 a_1 + \cdots + \lambda_n a_n \mid n \in \mathbb{N} \setminus \{0\}, \lambda_1, \ldots, \lambda_n \in \mathbb{N} \setminus \{0\}, a_1, \ldots, a_n \in A \}$$

(where $\mathbb{N}$ denotes the set of nonnegative integers). We say that *S* is generated by $A \subseteq S$ if $S = \langle A \rangle$. In this case, *A* is a *system of generators* of *S*. If *A* has finitely many elements, then we say that *S* is finitely generated.

A semigroup *M* is a *monoid* if it has an identity element, that is, there is an element in *M*, denoted by 0, such that $0 + a = a + 0 = a$ for all $a \in M$ (recall that we are assuming that the semigroups in this book are commutative, whence this also extends to monoids).

A subset *N* of *M* is a *submonoid* of *M* if it is a subsemigroup of *M* and $0 \in N$. Observe if *M* is a monoid, then $\{0\}$ is a submonoid of *M*. This is called the *trivial* submonoid of *M*. As for semigroups, the intersection of submonoids of a monoid is again one of its submonoids. Given a monoid *M* and a subset *A* of *M*, the smallest submonoid of *M* containing *A* is

$$\langle A \rangle = \{ \lambda_1 a_1 + \cdots + \lambda_n a_n \mid n \in \mathbb{N}, \ \lambda_1, \ldots, \lambda_n \in \mathbb{N} \text{ and } a_1, \ldots, a_n \in A \},$$

which we will call the submonoid of *M* generated by *A*. As in the semigroup case, the set *A* is a *system of generators* of *M* if $\langle A \rangle = M$, and we will also say that *M* is generated by *A*. Accordingly, a monoid *M* is *finitely generated* if there exists a system of generators of *M* with finitely many elements. Note that $\langle \emptyset \rangle = \{0\} = \langle 0 \rangle$.

Given two semigroups *X* and *Y*, a map $f : X \rightarrow Y$ is a *semigroup homomorphism* if $f(a + b) = f(a) + f(b)$ for all $a, b \in X$. We say that *f* is a *monomorphism*, an *epimorphism*, or an *isomorphism* if *f* is injective, surjective or bijective, respectively. Clearly, if *f* is an isomorphism so is its inverse f^{-1}. Two semigroups *X* and *Y* are said to be isomorphic if there exists an isomorphism between them. We will denote this fact by $X \cong Y$.

A map $f : X \rightarrow Y$ with *X* and *Y* monoids is a *monoid homomorphism* if it is a semigroup homomorphism and $f(0) = 0$. The concepts of monomorphism, epimorphism, and isomorphism of monoids are defined as for semigroups.

2 Multiplicity and embedding dimension

The set $\mathbb{N}$ with the operation of addition is a monoid. In this book we are mainly interested in the submonoids of $\mathbb{N}$. We see next that they can be classified up to isomorphism by those having finite complement in $\mathbb{N}$. A submonoid of $\mathbb{N}$ with finite complement in $\mathbb{N}$ is a *numerical semigroup*. In this section we show that every numerical semigroup (and thus every submonoid of $\mathbb{N}$) is finitely generated, admits a unique minimal system of generators and its cardinality is upper bounded by the least positive element in the monoid.

For A a nonempty subset of $\mathbb{N}$, $\langle A \rangle$, the submonoid of $\mathbb{N}$ generated by A, is a numerical semigroup if and only if the greatest common divisor of the elements of A is one.

Lemma 2.1. *Let A be a nonempty subset of $\mathbb{N}$. Then $\langle A \rangle$ is a numerical semigroup if and only if $\gcd(A) = 1$.*

Proof. Let $d = \gcd(A)$. Clearly, if s belongs to $\langle A \rangle$, then $d \mid s$. As $\langle A \rangle$ is a numerical semigroup, $\mathbb{N} \setminus \langle A \rangle$ is finite, and thus there exists a positive integer x such that $d \mid x$ and $d \mid x+1$. This forces d to be one.

For the converse, it suffices to prove that $\mathbb{N} \setminus \langle A \rangle$ is finite. Since $1 = \gcd(A)$, there exist integers $z_1, \ldots, z_n$ and $a_1, \ldots, a_n \in A$ such that $z_1 a_1 + \cdots + z_n a_n = 1$. By moving those terms with z_i negative to the right-hand side, we can find $i_1, \ldots, i_k, j_1, \ldots, j_l \in \{1, \ldots, n\}$ such that $z_{i_1} a_{i_1} + \cdots + z_{i_k} a_{i_k} = 1 - z_{j_1} a_{j_1} - \cdots - z_{j_l} a_{j_l}$. Hence there exists $s \in \langle A \rangle$ such that $s+1$ also belongs to $\langle A \rangle$. We prove that if $n \geq (s-1)s + (s-1)$, then $n \in \langle A \rangle$. Let q and r be integers such that $n = qs + r$ with $0 \leq r < s$. From $n \geq (s-1)s + (s-1)$, we deduce that $q \geq s - 1 \geq r$. It follows that $n = (rs + r) + (q-r)s = r(s+1) + (q-r)s \in \langle A \rangle$. $\qquad\square$

Numerical semigroups classify, up to isomorphism, the set of submonoids of $\mathbb{N}$.

Proposition 2.2. *Let M be a nontrivial submonoid of $\mathbb{N}$. Then M is isomorphic to a numerical semigroup.*

Proof. Let $d = \gcd(M)$. By Lemma 2.1, we know that $S = \langle \{ \frac{m}{d} \mid m \in M \} \rangle$ is a numerical semigroup. The map $f : M \to S$, $f(m) = \frac{m}{d}$ is clearly a monoid isomorphism. $\qquad\square$

If A and B are subsets of integer numbers, we write $A + B = \{a + b \mid a \in A, b \in B\}$. Thus for a numerical semigroup S, if we write $S^* = S \setminus \{0\}$, the set $S^* + S^*$ corresponds with those elements in S that are the sum of two nonzero elements in S.

Lemma 2.3. *Let S be a submonoid of $\mathbb{N}$. Then $S^* \setminus (S^* + S^*)$ is a system of generators of S. Moreover, every system of generators of S contains $S^* \setminus (S^* + S^*)$.*

Proof. Let s be an element of S^*. If $s \notin S^* \setminus (S^* + S^*)$, then there exist $x, y \in S^*$ such that $s = x + y$. We repeat this procedure for x and y, and after a finite number of steps $(x, y < s)$ we find $s_1, \ldots, s_n \in S^* \setminus (S^* + S^*)$ such that $s = s_1 + \cdots + s_n$. This proves that $S^* \setminus (S^* + S^*)$ is a system of generators of S.

Now, let A be a system of generators of S. Take $x \in S^* \setminus (S^* + S^*)$. There exist $n \in \mathbb{N} \setminus \{0\}$, $\lambda_1, \ldots, \lambda_n \in \mathbb{N}$ and $a_1, \ldots, a_n \in A$ such that $x = \lambda_1 a_1 + \cdots + \lambda_n a_n$. As $x \notin S^* + S^*$, we deduce that $x = a_i$ for some $i \in \{1, \ldots, n\}$. $\qquad\square$

This property also holds for any submonoid S of $\mathbb{N}^r$ for any positive integer r. The idea is that whenever $s = x + y$ with x and y nonzero, then x is strictly less than s with the usual partial order on $\mathbb{N}^r$. And there are only finitely many elements $x \in \mathbb{N}^r$

with $x \leq s$. However the set $S^* \setminus (S^* + S^*)$ needs not be finite for r greater than one (see Exercise 2.15). We are going to see that for $r = 1$ this set is always finite. To this end we introduce what is probably the most versatile tool in numerical semigroup theory.

Let S be a numerical semigroup and let n be one of its nonzero elements. The Apéry set (named so in honour of [2]) of n in S is

$$\text{Ap}(S,n) = \{s \in S \mid s - n \notin S\}.$$

Lemma 2.4. *Let S be a numerical semigroup and let n be a nonzero element of S. Then $\text{Ap}(S,n) = \{0 = w(0), w(1), \ldots, w(n-1)\}$, where $w(i)$ is the least element of S congruent with i modulo n, for all $i \in \{0, \ldots, n-1\}$.*

Proof. It suffices to point out that for every $i \in \{1, \ldots, n-1\}$, there exists $k \in \mathbb{N}$ such that $i + kn \in S$. $\square$

Example 2.5. Let S be the numerical semigroup generated by $\{5,7,9\}$. Then $S = \{0,5,7,9,10,12,14,\rightarrow\}$ (the symbol $\rightarrow$ means that every integer greater than 14 belongs to the set). Hence $\text{Ap}(S,5) = \{0,7,9,16,18\}$.

Observe that the above lemma in particular implies that the cardinality of $\text{Ap}(S,n)$ is n. With this result, we easily deduce the following.

Lemma 2.6. *Let S be a numerical semigroup and let $n \in S \setminus \{0\}$. Then for all $s \in S$, there exists a unique $(k,w) \in \mathbb{N} \times \text{Ap}(S,n)$ such that*

$$s = kn + w.$$

This lemma does not hold for submonoids of $\mathbb{N}^r$ in general. However, there are certain families of submonoids of $\mathbb{N}^r$ for which a similar property holds, and this apparently innocuous result makes it possible to translate some of the known results for numerical results to a more general scope (see [78]).

We say that a system of generators of a numerical semigroup is a *minimal* system of generators if none of its proper subsets generates the numerical semigroup.

Theorem 2.7. *Every numerical semigroup admits a unique minimal system of generators. This minimal system of generators is finite.*

Proof. Lemma 2.3 states that $S^* \setminus (S^* + S^*)$ is the minimal system of generators of S. By Lemma 2.6, we have that for any $n \in S^*$, we get that $S = \langle \text{Ap}(S,n) \cup \{n\} \rangle$. As $\text{Ap}(S,n) \cup \{n\}$ is finite, we deduce that $S^* \setminus (S^* + S^*)$ is finite. $\square$

As every submonoid of $\mathbb{N}$ is isomorphic to a numerical semigroup, this property translates to submonoids of $\mathbb{N}$.

Corollary 2.8. *Let M be a submonoid of $\mathbb{N}$. Then M has a unique minimal system of generators, which in addition is finite.*

Proof. Set $d = \gcd(M)$. Then $T = \left\{ \frac{x}{d} \mid x \in M \right\}$ is a submonoid of $\mathbb{N}$ such that $\gcd(T) = 1$. In view of Lemma 2.1, this means that T is a numerical semigroup. If A is the minimal system of generators of T, then $\{ da \mid a \in A \}$ is the minimal system of generators of M. $\qquad\square$

Corollary 2.9. *Let M be a submonoid of $\mathbb{N}$ generated by $\{0 \neq m_1 < m_2 < \cdots < m_p\}$. Then $\{m_1, \ldots, m_p\}$ is a minimal system of generators of M if and only if $m_{i+1} \notin \langle m_1, \ldots, m_i \rangle$.*

Let S be a numerical semigroup and let $\{n_1 < n_2 < \cdots < n_p\}$ be its minimal system of generators. Then n_1 is known as the *multiplicity* of S, denoted by $\mathrm{m}(S)$. The cardinality of the minimal system of generators, p, is called the *embedding dimension* of S and will be denoted by $\mathrm{e}(S)$.

Proposition 2.10. *Let S be a numerical semigroup. Then*

1) $\mathrm{m}(S) = \min(S \setminus \{0\})$,
2) $\mathrm{e}(S) \leq \mathrm{m}(S)$.

Proof. Clearly the multiplicity is the least positive integer in S. The other statement follows from the fact that $\{\mathrm{m}(S)\} \cup \mathrm{Ap}(S, \mathrm{m}(S)) \setminus \{0\}$ is a system of generators of S with cardinality $\mathrm{m}(S)$. $\qquad\square$

Observe that $\mathrm{e}(S) = 1$ if and only if $S = \mathbb{N}$. If m is a positive integer, then clearly $S = \{0, m, \rightarrow\}$ is a numerical semigroup with multiplicity m. It is easy to check that a minimal system of generators for S is $\{m, m+1, \ldots, 2m-1\}$. Hence $\mathrm{e}(S) = m = \mathrm{m}(S)$.

3 Frobenius number and genus

Frobenius in his lectures proposed the problem of giving a formula for the largest integer that is not representable as a linear combination with nonnegative integer coefficients of a given set of positive integers whose greatest common divisor is one. He also threw the question of determining how many positive integers do not have such a representation. By using our terminology, the first problem is equivalent to give a formula, in terms of the elements in a minimal system of generators of a numerical semigroup S, for the greatest integer not in S. This element is usually known as the *Frobenius number* of S, though in the literature it is sometimes replaced by the *conductor* of S, which is the least integer x such that $x + n \in S$ for all $n \in \mathbb{N}$. The Frobenius number of S is denoted here by $\mathrm{F}(S)$ and it is the conductor of S minus one. As for the second problem, the set of elements in $\mathrm{G}(S) = \mathbb{N} \setminus S$ is known as the set of *gaps* of S. Its cardinality is the *genus* of S, $\mathrm{g}(S)$, which is sometimes referred to as the *degree of singularity* of S.

Example 2.11. Let $S = \langle 5, 7, 9 \rangle$. We know that $S = \{0, 5, 7, 9, 10, 12, 14, \rightarrow\}$ and thus $\mathrm{F}(S) = 13$, $\mathrm{G}(S) = \{1, 2, 3, 4, 6, 8, 11, 13\}$ and $\mathrm{g}(S) = 8$.

There is no known general formula for the Frobenius number nor for the genus for numerical semigroups with embedding dimension greater than two (see [21] where it is shown that no polynomial formula can be found in this setting or the monograph [52] for a state of the art of the problem). However, if the Apéry set of any nonzero element of the semigroup is known, then both invariants are easy to compute.

Proposition 2.12 ([101]). *Let S be a numerical semigroup and let n be a nonzero element of S. Then*

1) $F(S) = (\max \mathrm{Ap}\,(S,n)) - n$,
2) $g(S) = \frac{1}{n}(\sum_{w \in \mathrm{Ap}(S,n)} w) - \frac{n-1}{2}$.

Proof. Note that by the definition of the elements in the Apéry set, $(\max \mathrm{Ap}\,(S,n)) - n \notin S$. If $x > (\max \mathrm{Ap}\,(S,n)) - n$, then $x + n > \max \mathrm{Ap}(S,n)$. Let $w \in \mathrm{Ap}(S,n)$ be such that w and $x + n$ are congruent modulo n. As $w < x + n$, this implies that $x = w + kn$ for some positive integer k, and consequently $x - n = w + (k-1)n$ belongs to S.

Observe that for every $w \in \mathrm{Ap}(S,n)$, if w is congruent with i modulo n and $i \in \{0, \ldots, n-1\}$, then there exists a nonnegative integer k_i such that $w = k_i n + i$. Thus, by using the notation of Lemma 2.4,

$$\mathrm{Ap}\,(S,n) = \{0, w(1) = k_1 n + 1, w(2) = k_2 n + 2, \ldots, w(n-1) = k_{n-1}n + n - 1\}.$$

An integer x congruent with $w(i)$ modulo n belongs to S if and only if $w(i) \leq x$. Thus

$$\begin{aligned} g(S) &= k_1 + \cdots + k_{n-1} \\ &= \frac{1}{n}((k_1 n + 1) + \cdots + (k_{n-1}n + n - 1)) - \frac{n-1}{2} \\ &= \frac{1}{n} \sum_{w \in \mathrm{Ap}(S,n)} w - \frac{n-1}{2}. \end{aligned}$$ $\square$

If S is a numerical semigroup minimally generated by $\langle a, b \rangle$, then

$$\mathrm{Ap}\,(S,a) = \{0, b, 2b, \ldots, (a-1)b\}$$

and Proposition 2.12 tells us the following result that goes back to the end of the 19th century.

Proposition 2.13 ([102]). *Let a and b be positive integers with $\gcd(a,b) = 1$.*

1) $F(\langle a, b \rangle) = ab - a - b$,
2) $g(\langle a, b \rangle) = \frac{ab-a-b+1}{2}$.

Observe that for numerical semigroups of embedding dimension two $g(S) = (F(S) + 1)/2$ (and thus $F(S)$ is always an odd integer). This is not in general the case for higher embedding dimensions, though this property characterizes a very interesting class of numerical semigroups as we will see later.

If S is a numerical semigroup and $s \in S$, then $F(S) - s$ cannot be in S. From this we obtain that the above equality is just an inequality in general.

Lemma 2.14. *Let S be a numerical semigroup. Then*

$$g(S) \geq \frac{F(S) + 1}{2}.$$

Thus numerical semigroups for which the equality holds are numerical semigroups with the "least" possible number of gaps.

Remark 2.15. If one fixes a positive integer f, then it is not true in general that there are more numerical semigroups with Frobenius number $f + 1$ than numerical semigroups with Frobenius number f. The following table can be found in [91] ($\mathrm{ns}(F)$ stands for the number of numerical semigroups with Frobenius number F).

F	$\mathrm{ns}(F)$	F	$\mathrm{ns}(F)$	F	$\mathrm{ns}(F)$
1	1	14	103	27	16132
2	1	15	200	28	16267
3	2	16	205	29	34903
4	2	17	465	30	31822
5	5	18	405	31	70854
6	4	19	961	32	68681
7	11	20	900	33	137391
8	10	21	1828	34	140661
9	21	22	1913	35	292081
10	22	23	4096	36	270258
11	51	24	3578	37	591443
12	40	25	8273	38	582453
13	106	26	8175	39	1156012

Bras-Amorós in [10] has computed the number of numerical semigroups with genus g for $g \in \{0, \ldots, 50\}$, and her computations show a Fibonacci like behavior on the number of numerical semigroups with fixed genus less than or equal to 50. However it is still not known in general if for a fixed positive integer g there are more numerical semigroups with genus $g + 1$ than numerical semigroups with genus g. We reproduce in Table 1 the results obtained by Bras-Amorós (in the table n_g stands for the number of numerical semigroups with genus g).

Lemma 2.16. *Let S be a numerical semigroup generated by $\{n_1, n_2, \ldots, n_p\}$. Let $d = \gcd\{n_1, \ldots, n_{p-1}\}$ and set $T = \langle n_1/d, \ldots, n_{p-1}/d, n_p \rangle$. Then*

$$\mathrm{Ap}(S, n_p) = d(\mathrm{Ap}(T, n_p)).$$

Table 1 Number of numerical semigroups of genus up 50.

g	n_g	$n_{g-1}+n_{g-2}$	$(n_{g-1}+n_{g-2})/n_g$	n_g/n_{g-1}
0	1			
1	1			1
2	2	2	1	2
3	4	3	0.75	2
4	7	6	0.857143	1.75
5	12	11	0.916667	1.71429
6	23	19	0.826087	1.91667
7	39	35	0.897436	1.69565
8	67	62	0.925373	1.71795
9	118	106	0.898305	1.76119
10	204	185	0.906863	1.72881
11	343	322	0.938776	1.68137
12	592	547	0.923986	1.72595
13	1001	935	0.934066	1.69088
14	1693	1593	0.940933	1.69131
15	2857	2694	0.942947	1.68754
16	4806	4550	0.946733	1.68218
17	8045	7663	0.952517	1.67395
18	13467	12851	0.954259	1.67396
19	22464	21512	0.957621	1.66808
20	37396	35931	0.960825	1.66471
21	62194	59860	0.962472	1.66312
22	103246	99590	0.964589	1.66006
23	170963	165440	0.967695	1.65588
24	282828	274209	0.969526	1.65432
25	467224	453791	0.971249	1.65197
26	770832	750052	0.973042	1.64981
27	1270267	1238056	0.974642	1.64792
28	2091030	2041099	0.976121	1.64613
29	3437839	3361297	0.977735	1.64409
30	5646773	5528869	0.97912	1.64254
31	9266788	9084612	0.980341	1.64108
32	15195070	14913561	0.981474	1.63973
33	24896206	24461858	0.982554	1.63844
34	40761087	40091276	0.983567	1.63724
35	66687201	65657293	0.984556	1.63605
36	109032500	107448288	0.98547	1.63498
37	178158289	175719701	0.986312	1.63399
38	290939807	287190789	0.987114	1.63304
39	474851445	469098096	0.987884	1.63213
40	774614284	765791252	0.98861	1.63128
41	1262992840	1249465729	0.98929	1.63048
42	2058356522	2037607124	0.989919	1.62975
43	3353191846	3321349362	0.990504	1.62906
44	5460401576	5411548368	0.991053	1.62842
45	8888486816	8813593422	0.991574	1.62781
46	14463633648	14348888392	0.992067	1.62723
47	23527845502	23352120464	0.992531	1.62669
48	38260496374	37991479150	0.992969	1.62618
49	62200036752	61788341876	0.993381	1.6257
50	101090300128	100460533126	0.99377	1.62525

Proof. If $w \in \mathrm{Ap}\,(S, n_p)$, then $w \in \langle n_1, \ldots, n_{p-1} \rangle$. Hence $w/d \in \langle n_1/d, \ldots, n_{p-1}/d \rangle \subseteq T$. If $w/d - n_p \in T$, then $w - dn_p \in S$, which is impossible.

Now take $w \in \mathrm{Ap}\,(T, n_p)$. Then $w \in \langle n_1/d, \ldots, n_{p-1}/d \rangle$, and thus $dw \in \langle n_1, \ldots, n_{p-1} \rangle \subseteq S$. If $dw - n_p$ also belongs to S, then $dw - n_p = \lambda_1 n_1 + \cdots + \lambda_{p-1} n_{p-1} + \lambda_p n_p$ for some $\lambda_1, \ldots, \lambda_p \in \mathbb{N}$. Since S is a numerical semigroup gcd$\{n_1, \ldots, n_p\} = 1$, which implies that gcd$\{d, n_p\} = 1$. This leads to $d|(\lambda_p + 1)$, because $(\lambda_p + 1)n_p = dw - (\lambda_1 n_1 + \cdots + \lambda_{p-1} n_{p-1})$. But then $w = \frac{\lambda_1 n_1}{d} + \cdots + \frac{\lambda_{p-1} n_{p-1}}{d} + \frac{\lambda_p + 1}{d} n_p$, with $(\lambda_p + 1)/d$ a positive integer, contradicting that $w \in \mathrm{Ap}\,(T, n_p)$. $\square$

By putting Proposition 2.12 and Lemma 2.16 together, we obtain the following property.

Proposition 2.17 ([42]). *Let S be a numerical semigroup with minimal system of generators $\{n_1, n_2, \ldots, n_p\}$. Let $d = \gcd\{n_1, \ldots, n_{p-1}\}$ and set $T = \langle n_1/d, \ldots, n_{p-1}/d, n_p \rangle$. Then*

1) $\mathrm{F}(S) = d\mathrm{F}(T) + (d - 1)n_p$,
2) $\mathrm{g}(S) = d\mathrm{g}(T) + \frac{(d-1)(n_p-1)}{2}$.

Example 2.18. Let $S = \langle 20, 30, 17 \rangle$. As $\gcd\{20, 30\} = 10$, $T = \langle 2, 3, 17 \rangle = \langle 2, 3 \rangle$. Hence $\mathrm{F}(S) = 10\mathrm{F}(T) + 9 \cdot 17 = 10 + 153 = 163$, and $\mathrm{g}(S) = 10\mathrm{g}(T) + \frac{9 \cdot 16}{2} = 10 + 72 = 82$.

4 Pseudo-Frobenius numbers

Let S be a numerical semigroup. Following the notation introduced in [71], we say that an integer x is a *pseudo-Frobenius number* if $x \notin S$ and $x + s \in S$ for all $s \in S \setminus \{0\}$. We will denote by $\mathrm{PF}(S)$ the set of pseudo-Frobenius numbers of S, and its cardinality, which deserves a name of its own, is the *type* of S, denoted by $\mathrm{t}(S)$. From the definition it easily follows that $\mathrm{F}(S) \in \mathrm{PF}(S)$, in fact it is the maximum of this set.

Over the set of integers we can define the following relation: $a \leq_S b$ if $b - a \in S$. As S is a numerical semigroup, it easily follows that this relation is an order relation (reflexive, transitive and antisymmetric). From the definition of pseudo-Frobenius numbers, we obtain that they are the maximal elements with respect to $\leq_S$ of $\mathbb{Z} \setminus S$ ($\mathbb{Z}$ denotes the set of integers).

Proposition 2.19. *Let S be a numerical semigroup. Then*

1) $\mathrm{PF}(S) = \mathrm{Maximals}_{\leq_S}(\mathbb{Z} \setminus S)$,
2) $x \in \mathbb{Z} \setminus S$ if and only if $f - x \in S$ for some $f \in \mathrm{PF}(S)$.

This result establishes a sort of duality between minimal generators and pseudo-Frobenius numbers of a numerical semigroup, since $\mathrm{Minimals}_{\leq_S}(S \setminus \{0\})$ is the minimal system of generators of S.

A characterization in terms of the Apéry sets already appears in [32, Proposition 7].

Proposition 2.20. *Let S be a numerical semigroup and let n be a nonzero element of S. Then*

$$\mathrm{PF}(S) = \{\, w - n \mid w \in \mathrm{Maximals}_{\leq_S} \mathrm{Ap}\,(S,n)\,\}.$$

Proof. Let $x \in \mathrm{PF}(S)$. Hence $x \notin S$ and $x + n \in S$, or in other words, $x + n \in \mathrm{Ap}\,(S,n)$. Let $w \in \mathrm{Ap}\,(S,n)$ be such that $x + n \leq_S w$. Then $w - (x+n) = w - n - x \in S$. This means that $w - n = x + s$ for some $s \in S$. As $w - n \notin S$ and $x \in \mathrm{PF}(S)$, this forces s to be zero and thus $w = x + n$.

Now take $w \in \mathrm{Maximals}_{\leq_S} \mathrm{Ap}\,(S,n)$. Then $w - n \notin S$. If $w - n + s \notin S$ for some nonzero element s of S, then $w + s \in \mathrm{Ap}\,(S,n)$, contradicting the maximality of w.   □

Example 2.21. Let $S = \langle 5,7,9 \rangle$. Then $\mathrm{Maximals}_{\leq_S}\mathrm{Ap}\,(S,5) = \{16,18\}$. Hence $\mathrm{PF}(S) = \{11,13\}$.

Example 2.22. If S is a numerical semigroup minimally generated by $\langle a,b \rangle$, then as we have pointed out above,

$$\mathrm{Ap}\,(S,a) = \{0, b, 2b, \ldots, (a-1)b\}.$$

This implies that $\mathrm{Maximals}_{\leq_S}\mathrm{Ap}\,(S,a) = \{(a-1)b\}$ and $\mathrm{PF}(S) = \{ab - a - b\}$. Thus numerical semigroups with embedding dimension two have type one.

As the cardinality of $\mathrm{Ap}\,(S,n)$ is n and the zero element is never a maximal element, from the above proposition we obtain an upper bound for the type of a numerical semigroup.

Corollary 2.23. *Let S be a numerical semigroup. Then*

$$\mathrm{t}(S) \leq \mathrm{m}(S) - 1.$$

Recall that the embedding dimension of a numerical semigroup does not exceed the multiplicity of the numerical semigroup. We already know (Example 2.22) that numerical semigroups with embedding dimension two have type one. We will see in Chapter 9 that numerical semigroups with embedding dimension three have type one or two. However, for embedding dimension greater than three, the type is not upper bounded by the embedding dimension as the following example due to Backelin shows.

Example 2.24. [32] Let $S = \langle s, s+3, s+3n+1, s+3n+2 \rangle$. For $n \geq 2$, $r \geq 3n+2$ and $s = r(3n+2) + 3$, the type of S is $2n + 3$.

Example 2.25. Let m be an integer greater than one. Note that for $S = \{0, m, \rightarrow\}$, $\mathrm{PF}(S) = \{1, 2, \ldots, m-1\}$. These semigroups reach the bound given in the above corollary.

Let S be a numerical semigroup. Denote by

$$\mathrm{N}(S) = \{\, s \in S \mid s < \mathrm{F}(S)\,\}.$$

This set fully determines S. Its cardinality is denoted by $n(S)$. Clearly $g(S)+n(S) = F(S)+1$.

We already know that if x is an integer not in S, there exists $f \in PF(S)$ such that $x \leq_S f$. Define $f_x = \min\{f \in PF(S) \mid f-x \in S\}$. Then the map

$$G(S) \to PF(S) \times N(S),\ x \mapsto (f_x, f_x - x)$$

is injective, which proves the following bound.

Proposition 2.26. [32, Theorem 20] *Let S be a numerical semigroup. Then*

$$g(S) \leq t(S)n(S).$$

This inequality is equivalent to $F(S)+1 \leq (t(S)+1)n(S)$. Wilf in [108] conjectured that $F(S)+1 \leq e(S)n(S)$. For some families of numerical semigroups this conjecture is known to be true, but the general case remains unsolved.

Example 2.27. The following picture is obtained by using the GAP package numericalsgps ([23]). The picture represents the Apéry set of 11 in the semigroup S ordered by the relation $\leq_S$. We also illustrate the use of some functions related to the elements described in this chapter.

```
gap> S:=NumericalSemigroup( 11, 12, 13, 32, 53 );;
gap> MinimalGeneratingSystemOfNumericalSemigroup(S);
[ 11, 12, 13, 32, 53 ]
gap> MultiplicityOfNumericalSemigroup(S);
11
gap> FrobeniusNumberOfNumericalSemigroup(S);
42
gap> GapsOfNumericalSemigroup(S);
[ 1, 2, 3, 4, 5, 6, 7, 8, 9, 10, 14, 15, 16, 17,
18, 19, 20, 21, 27, 28, 29, 30, 31, 40, 41, 42 ]
gap> DrawAperyListOfNumericalSemigroup(
> AperyListOfNumericalSemigroupWRTElement(S,11));
```

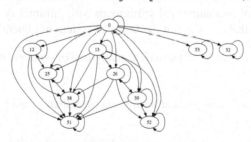

```
gap> PseudoFrobeniusOfNumericalSemigroup(S);
[ 21, 40, 41, 42 ]
gap> SmallElementsOfNumericalSemigroup(S);
[ 0, 11, 12, 13, 22, 23, 24, 25, 26, 32, 33,
34, 35, 36, 37, 38, 39, 43 ]
```

Example 2.28. And now an example of a numerical semigroup with less embedding dimension than type.

```
gap> S:=NumericalSemigroup(41,152,373,407);
<Numerical semigroup with 4 generators>
gap> MinimalGeneratingSystemOfNumericalSemigroup(S);
[ 41, 152, 373, 407 ]
gap> PseudoFrobeniusOfNumericalSemigroup(S);
[ 1161, 1195, 1332, 1381, 1451, 1479, 1582 ]
```

Exercises

Exercise 2.1. Let S be a numerical semigroup and let x be an element of S. Prove that $S \setminus \{x\}$ is a numerical semigroup if and only if x belongs to the minimal system of generators of S.

Exercise 2.2. Prove that the intersection of finitely many numerical semigroups is a numerical semigroup. Show with an example that the result does not hold for infinite intersections.

Exercise 2.3. Let S and T be two numerical semigroups. Take $m \in S \cap T$ with $m \neq 0$. Prove that if $\mathrm{Ap}(S,m) = \{0, u_1, \ldots, u_{m-1}\}$ and $\mathrm{Ap}(T,m) = \{0, v_1, \ldots, v_{m-1}\}$ (with u_i and v_i the smallest elements congruent with i modulo m in S and T, respectively), then $\mathrm{Ap}(S \cap T, m) = \{0, \max\{u_1, v_1\}, \ldots, \max\{u_{m-1}, v_{m-1}\}\}$.

Exercise 2.4. Let S and T be two numerical semigroups. Prove that

a) $S + T$ is a numerical semigroup,
b) $S + T$ is the smallest numerical semigroup containing $S \cup T$,
c) if A and B are systems of generators of S and T, respectively, then $A \cup B$ is a system of generators of $S + T$.

Give an example of two numerical semigroups with minimal systems of generators A and B and such that $A \cup B$ is not the minimal system of generators of $\langle A \cup B \rangle$.

Exercise 2.5. Prove that S is a numerical semigroup of type one if and only if $\mathrm{g}(S) = \frac{\mathrm{F}(S)+1}{2}$.

Exercise 2.6. Let S be a numerical semigroup other than $\mathbb{N}$ and assume that $\mathrm{PF}(S) = \{f_1 > f_2 > \cdots > f_t\}$.

a) Prove that $S \cup \{f_1, \ldots, f_k\}$ is a numerical semigroup for all $k \in \{1, \ldots, t\}$.
b) Prove that if $\mathrm{F}(S) > \mathrm{m}(S)$, then $T = S \cup \{\mathrm{F}(S)\}$ is a numerical semigroup with $\mathrm{e}(S) \leq \mathrm{e}(T) \leq \mathrm{e}(S) + 1$.
c) Give examples in which $\mathrm{e}(T) = \mathrm{e}(S)$ and $\mathrm{e}(S) + 1 = \mathrm{e}(T)$.

Exercise 2.7. Let a and b be two integers with $0 < b < a$, and let $S = \langle a, a + 1, \ldots, a + b \rangle$. Prove that $F(S) = \lceil \frac{a-1}{b} \rceil a - 1$, where for q a rational number, $\lceil q \rceil$ denotes the minimum of the integers greater than or equal to q (see [33] for the computation of other notable elements of these semigroups).

Exercise 2.8 ([8]). Let S be a numerical semigroup with positive conductor c. Let $d = \max\{s \in S \mid s < c\}$. This integer is the *dominant* of S.

a) Prove that the dominant of S is zero if and only if $S = \{0, c, \rightarrow\}$. These numerical semigroups are sometimes called in the literature *half-lines* or *ordinary*.

b) Assume that S has positive dominant d. For $s \in S$, define $g(s)$ as the cardinality of $\{x \in G(S) \mid x < s\}$ (note that $g(d) < g(c) = g(S)$). Set

$$c' = \min\{s \in S \mid g(s) = g(d)\}, \quad d' = \max\{s \in S \mid s < c'\}.$$

We say that S is *acute* if $c - d \leq c' - d'$. Prove that every numerical semigroup of the form $\langle a, a+1, \ldots, a+b \rangle$ that is not a half line (or equivalently, with a and b positive integers such that $b < a - 1$) is acute.

Exercise 2.9. Let S be a numerical semigroup and let m be a nonzero element of S. Show that $T = (m + S) \cup \{0\}$ is also a numerical semigroup with $m(T) = m$, $e(T) = m$ and $t(T) = m - 1$.

Exercise 2.10. Prove that Wilf's conjecture holds in the following cases.

a) For every numerical semigroup of type 1.
b) For every numerical semigroup of type 2.
c) For numerical semigroups S with $e(S) = m(S) - 1$.

Exercise 2.11 ([63]). Prove that if S is a numerical semigroup with multiplicity three, then

$$S = \langle 3, 3g(S) - F(S), F(S) + 3 \rangle.$$

Exercise 2.12 ([63]). Let S be a numerical semigroup with minimal system of generators $\{n_1 < n_2 < \cdots < n_p\}$. Then we say that n_2 is the *ratio* of S. Prove that if S is a numerical semigroup with multiplicity four and ratio r, then

$$S = \langle 4, r, 4g(S) - F(S) - r + 2, F(S) + 4 \rangle.$$

Exercise 2.13. Let S be a numerical semigroup. A *relative ideal* I of S is a subset of $\mathbb{Z}$ such that $I + S \subseteq I$ and $s + I = \{s + i \mid i \in I\} \subset S$ for some $s \in S$. An *ideal* of S is a relative ideal of S contained in S.

a) If H and K are relative ideals of S, then prove that

$$H - K = \{z \in \mathbb{Z} \mid z + K \subseteq H\}$$

is also a relative ideal of S.

b) $M = S^*$ is the maximal (with respect to set inclusion) ideal of S.

c) Given a relative ideal I of S, denote by $I^{\bullet} = S - I$. Prove that

   i) $I \subseteq I^{\bullet\bullet}$,
   ii) $I^{\bullet} = I^{\bullet\bullet\bullet}$.

d) Show that $\mathrm{PF}(S) = S \setminus M^{\bullet}$.

Exercise 2.14. [4] Let S be a numerical semigroup. Define

$$\Omega = \{z \in \mathbb{Z} \mid \mathrm{F}(S) - z \notin S\}.$$

   i) Prove that $S \subseteq \Omega \subseteq \mathbb{N}$ and that Ω is a relative ideal of S. This ideal is called the *canonical ideal* of S.
   ii) Show that the map $\mathrm{PF}(S) \to \Omega \setminus (\Omega + M)$, $f \mapsto \mathrm{F}(S) - f$ is a one-to-one correspondence (where M is the maximal ideal of S, Exercise 2.13). As a consequence, $\mathrm{t}(S)$ is the cardinality of $\Omega \setminus (\Omega + M)$.

Exercise 2.15. Let $M = \{(x,y) \in \mathbb{N}^2 \mid x > 5\} \cup \{(0,0)\}$. Prove that M is a submonoid of $\mathbb{N}^2$ that is not finitely generated.

Exercise 2.16. Let S be a numerical semigroup. The *enumeration* of S is the only increasing bijective map λ from $\mathbb{N}$ to S, that is, $\lambda(i)$ corresponds with the ith element of S. For $i \in \mathbb{N}$, let v_i denote the cardinality of $\{j \in \mathbb{N} \mid \lambda(i) - \lambda(j) \in S\}$. Define for i and j two nonnegative integers,

$$i \oplus j = \lambda^{-1}(\lambda(i) + \lambda(j)).$$

Prove that

a) the semigroup S is uniquely determined by the sequence $\{v_i\}_{i \in \mathbb{N}}$ ([8]),
b) for every nonnegative integer i, v_i is the number of pairs $(j,k) \in \mathbb{N}^2$ such that $j \oplus k = i$ (and thus it can be determined by the set of integers $j \oplus k$, with $j, k \in \{0, \ldots, i\}$, [9]).

Chapter 2
Numerical semigroups with maximal embedding dimension

Introduction

Even though the study and relevance of maximal embedding dimension numerical
semigroups arises in a natural way among the other numerical semigroups, they
have become specially renowned due to the existing applications to commutative
algebra via their associated semigroup ring (see for instance [1, 5, 15, 16, 99, 100]).
They are a source of examples of commutative rings with some maximal properties. As we mentioned in the introduction of Chapter 1, this is partially due to the
fact that the study of some attributes of an analytically unramified one-dimensional
local domains can be performed via their value semigroups. Of particular interest
are two subclasses of maximal embedding dimension numerical semigroups, which
are those semigroups having the Arf property and saturated numerical semigroups.
These two families are related with the problem of resolution of singularities in a
curve.

Inspired by [3], Lipman in [47] introduces and motivates the study of Arf rings.
The characterization of these rings via their value semigroups yields the Arf property for numerical semigroups. The reader can find in [5] a considerable amount
of characterizations of this property for numerical semigroups. Arf numerical semigroups have gained lately a particular interest due to their applications to algebraic
error correcting codes (see [18, 7] and the references given therein).

Saturated rings were introduced in three different ways by Zariski ([109]), Pham-
Teissier ([50]) and Campillo ([17]), though their definitions coincide for alge-
braically closed fields of zero characteristic. As for the Arf property, saturated numerical semigroups come into the scene after a characterization of saturated rings
in terms of their value semigroups (see [26, 49]).

J.C. Rosales, P.A. García-Sánchez, *Numerical Semigroups*,
Developments in Mathematics 20, DOI 10.1007/978-1-4419-0160-6_3,
© Springer Science+Business Media, LLC 2009

1 Characterizations

Let S be a numerical semigroup. We know that its embedding dimension, $e(S)$, is less than or equal to its multiplicity, $m(S)$. We say that S has maximal embedding dimension if $e(S) = m(S)$. In this section we give several characterizations of this property in terms of the notable elements presented in Chapter 1.

If x is a minimal generator of S, and $n \in S \setminus \{0, x\}$, then $x - n$ does not belong to S. This implies that $x \in \mathrm{Ap}(S, n)$.

Proposition 3.1. *Let S be a numerical semigroup minimally generated by $\{n_1 < n_2 < \cdots < n_e\}$. Then S has maximal embedding dimension if and only if $\mathrm{Ap}(S, n_1) = \{0, n_2, \ldots, n_e\}$.*

Proof. As we have pointed out above, $\{n_2, \ldots, n_e\} \subseteq \mathrm{Ap}(S, n_1) \setminus \{0\}$. We know that the cardinality of $\mathrm{Ap}(S, n_1)$ is n_1. Hence $e = n_1$ if and only if $\{0, n_2, \ldots, n_e\} = \mathrm{Ap}(S, n_1)$. $\square$

As a consequence of Propositions 2.12 and 2.20 we obtain the following properties.

Corollary 3.2. *Let S be a numerical semigroup minimally generated by $\{n_1 < n_2 < \cdots < n_e\}$.*

1) If S has maximal embedding dimension, then $\mathrm{F}(S) = n_e - n_1$.
2) S has maximal embedding dimension if and only if $g(S) = \frac{1}{n_1}(n_2 + \cdots + n_e) - \frac{n_1 - 1}{2}$.
3) S has maximal embedding dimension if and only if $t(S) = n_1 - 1$.

Example 3.3. The numerical semigroup $S = \langle 4, 5, 11 \rangle$ has $\mathrm{F}(S) = 11 - 4 = n_e - n_1$, but it does not have maximal embedding dimension.

Remark 3.4. Let S be a numerical semigroup minimally generated by $\{n_1 < n_2 < \cdots < n_e\}$.

1) We already know (Corollary 2.23) that $t(S) \leq m(S) - 1$. So the numerical semigroups with maximal embedding dimension are those with maximal type (in terms of their multiplicities).
2) As by Selmer's formula (Proposition 2.12) $g(S) \geq \frac{1}{n_1}(n_2 + \cdots + n_e) - \frac{n_1 - 1}{2}$, numerical semigroups with maximal embedding dimension can also be viewed as those with the least possible number of holes (in terms of their minimal generators).

Given a nonzero integer n and two integers a and b, we write $a \equiv b \bmod n$ to denote that n divides $a - b$. We denote by $b \bmod n$ the remainder of the division of b by n. The following result characterizes those subsets of positive integers that can be realized as Apéry sets of a numerical semigroup.

Proposition 3.5 ([57]). *Let $n \in \mathbb{N} \setminus \{0\}$ and let $C = \{w(0) = 0, w(1), \ldots, w(n - 1)\} \subseteq \mathbb{N}$ be such that $w(i)$ is congruent with i modulo n for all $i \in \{1, \ldots, n-1\}$. Let S be the numerical semigroup $\langle \{n\} \cup C \rangle$. The following conditions are equivalent.*

1) $\mathrm{Ap}(S, n) = C$.
2) For all $i, j \in \{1, \ldots, n-1\}$, $w(i) + w(j) \geq w((i + j) \bmod n)$.

Proof. Note that $w(i) + w(j)$ and $w((i + j) \bmod n)$ are congruent modulo n for all $i, j \in \{1, \ldots, n-1\}$. Hence Condition 2) is equivalent to

2')for all $i, j \in \{1, \ldots, n-1\}$, there exists $t \in \mathbb{N}$ such that $w(i) + w(j) = tn + w((i + j) \bmod n)$.

If $\mathrm{Ap}(S, n) = C$, then by Lemma 2.6, $w(i) + w(j) = kn + c$ for some $k \in \mathbb{N}$ and $c \in C$. Clearly $w(i) + w(j) \equiv c \bmod n$, and thus $c = w((i + j) \bmod n)$.

Now, assume that the second statement holds. Let us show that $\mathrm{Ap}(S, n) \subseteq C$. If $s \in \mathrm{Ap}(S, n) \subset S$, then there exist $c_1, \ldots, c_t \in C$ such that $s = \sum_{i=1}^{t} c_i$. By applying several times Condition 2'), we get that $s = kn + c$, with $c \in C$ and $k \in \mathbb{N}$. As $s \in \mathrm{Ap}(S, n)$, k must be zero and consequently $s = c \in C$.

In view of Lemma 2.4, the cardinality of $\mathrm{Ap}(S, n)$ is n. As the cardinality of C is also n and $\mathrm{Ap}(S, n) \subseteq C$, this forces $\mathrm{Ap}(S, n)$ to be equal to C. $\square$

As we have seen in Proposition 3.1, the Apéry sets of the multiplicity in numerical semigroups with maximal embedding dimension have special shapes. This together with the last characterization of Apéry sets yields an alternative way to distinguish numerical semigroups with maximal embedding dimension by looking at the Apéry sets of their multiplicities.

Corollary 3.6. *Let S be a numerical semigroup with multiplicity m and assume that $\mathrm{Ap}(S, m) = \{w(0) = 0, w(1), \ldots, w(m - 1)\}$ with $w(i) \equiv i \bmod m$ for all $i \in \{1, \ldots, m-1\}$. Then S has maximal embedding dimension if and only if for all $i, j \in \{1, \ldots, m-1\}$, $w(i) + w(j) > w((i + j) \bmod m)$.*

Proof. The necessity follows from Propositions 3.1 and 3.5.

By Lemma 2.6, we know that $S = \langle m, w(1), \ldots, w(m - 1) \rangle$. From the condition $w(i) + w(j) > w((i + j) \bmod m)$, we deduce that $\{m, w(1), \ldots, w(m - 1)\}$ is a minimal system of generators of S. Hence S has maximal embedding dimension. $\square$

Proposition 3.5 and Corollary 3.6 can be used to construct maximal embedding numerical semigroups from an arbitrary numerical semigroup.

Corollary 3.7. *Let S be a numerical semigroup and let n be a positive integer in S. Then $\langle n, w(1) + n, \ldots, w(n - 1) + n \rangle$ is a maximal embedding dimension numerical semigroup, where for all $i \in \{1, \ldots, n-1\}$, $w(i)$ is the element in $\mathrm{Ap}(S, n)$ congruent with i modulo n.*

Example 3.8. Let a and b be two positive integers greater than one with $\gcd\{a, b\} = 1$. We already know that $\mathrm{Ap}(\langle a, b \rangle, a) = \{0, b, 2b, \ldots, (a - 1)b\}$. By Corollary 3.7,

$$\langle a, a+b, a+2b, \ldots, a+(a-1)b \rangle$$

has maximal embedding dimension.

A sort of converse operation can be performed on a numerical semigroup with maximal embedding dimension. The proof also follows from Proposition 3.5 and Corollary 3.6.

Corollary 3.9 ([57]). *Let S be a numerical semigroup with maximal embedding dimension and multiplicity m. For all $i \in \{1, \ldots, m-1\}$, write $w(i)$ for the unique element in Ap (S, m) congruent with i modulo m. Define $T = \langle m, w(1) - m, \ldots, w(m-1) - m \rangle$. Then T is a numerical semigroup with Ap $(T, m) = \{0, w(1) - m, \ldots, w(m-1) - m\}$.*

From these two last results and Proposition 2.12, we obtain the following correspondence.

Corollary 3.10 ([57]). *There is a one to one correspondence between the set of numerical semigroups with multiplicity m and Frobenius number f, and the set of numerical semigroups with maximal embedding dimension, Frobenius number $f + m$, multiplicity m and the rest of minimal generators greater than 2m.*

Remark 3.11. If we want to construct the set of all numerical semigroups, according to this last result, it suffices to construct those having maximal embedding dimension. In other words, maximal embedding dimension numerical semigroups can be used to represent the whole class of numerical semigroups.

The following characterization can be deduced from [5, Proposition I.2.9]. For an integer z and a subset A of integers, the set $\{z + a \mid a \in A\}$ is denoted by $z + A$.

Proposition 3.12. *Let S be a numerical semigroup. The following conditions are equivalent.*

1) S has maximal embedding dimension.
2) For all $x, y \in S^*$, $x + y - \mathrm{m}(S) \in S$.
3) $-\mathrm{m}(S) + S^*$ is a numerical semigroup.

Proof. 1) implies 2). If either $x - \mathrm{m}(S) \in S$ or $y - \mathrm{m}(S) \in S$, then 2) follows trivially. So assume that both x and y are in Ap $(S, \mathrm{m}(S))$. The result now follows by Corollary 3.6.

2) implies 3). Trivial.

3) implies 1). Denote by $w(i)$ the unique element in Ap $(S, \mathrm{m}(S))$ congruent with i modulo m, $1 \le i \le m-1$. We use Corollary 3.6 again. If $w(i) + w(j) = w((i + j) \bmod \mathrm{m}(S))$ for some $i, j \in \{1, \ldots, \mathrm{m}(S) - 1\}$, then $w(i) - \mathrm{m}(S) + w(j) - \mathrm{m}(S) = w((i+j) \bmod \mathrm{m}(S)) - 2\mathrm{m}(S) \notin \{x - \mathrm{m}(S) \mid x \in S^*\}$, contradicting that this set is a numerical semigroup.   □

If in the last proposition we use T to denote the semigroup $-\mathrm{m}(S) + S^*$, then $S = (\mathrm{m}(S) + T) \cup \{0\}$. From this proposition it is not hard to prove the following characterization (see also Exercise 2.9).

Corollary 3.13 ([63]). *Let S be a numerical semigroup. Then S has maximal embedding dimension if and only if there exists a numerical semigroup T and $t \in T \setminus \{0\}$ such that $S = (t+T) \cup \{0\}$.*

Example 3.14. Let $S = \langle 4,5,7 \rangle = \{0,4,5,7,\rightarrow\}$. Then $T = (9+S) \cup \{0\} = \{0,9,13,14,16,\rightarrow\}$ is a maximal embedding dimension numerical semigroup. Note that $T = \langle 9,13,14,16,17,19,20,21,24 \rangle$.

Lemma 3.15. *Let S and T be numerical semigroups. Let $s \in S^*$ and $t \in T^*$. Then $(s+S) \cup \{0\} = (t+T) \cup \{0\}$ if and only if $S = T$ and $s = t$.*

Proof. Assume that $(s+S) \cup \{0\} = (t+T) \cup \{0\}$. Note that $\mathrm{m}((s+S) \cup \{0\}) = s$ and $\mathrm{m}((t+T) \cup \{0\}) = t$. Hence $s = t$. Moreover, $S = -s + (s+S) = -s + (t+T) = -t + (t+T) = T$. The other implication is trivial. $\qquad\square$

If S is a numerical semigroup and s is a nonzero element of S, then $s+S$ is an ideal of S (see Exercise 2.13). These ideals are called *principal* ideals of S. Numerical semigroups of the form $(x+S) \cup \{0\}$ with S a numerical semigroup and x a nonzero element of S are called in [63] *pi-semigroups* (where *pi* is an acronym of principal ideal). For a given numerical semigroup S define

$$\mathscr{P}I(S) = \{ (x+S) \cup \{0\} \mid x \in S^* \}.$$

If $x \neq 1$, it can be shown that $\mathrm{F}((x+S) \cup \{0\}) = \mathrm{F}(S) + x$ and that $\mathrm{g}((x+S) \cup \{0\}) = \mathrm{g}(S) + x - 1$ (clearly, the multiplicity of $(x+S) \cup \{0\}$ is x; see Exercises 2.9 and 3.6). From this it easily follows that two elements S_1 and S_2 in $\mathscr{P}I(S)$ coincide if and only if they have the same Frobenius number, or equivalently, they have the same genus.

Proposition 3.16. *The set $\{ \mathscr{P}I(S) \mid S$ is a numerical semigroup $\}$ is a partition of the set of numerical semigroups with maximal embedding dimension.*

Proof. Follows from Corollary 3.13 and Lemma 3.15. $\qquad\square$

This result is telling us that from a fixed numerical semigroup we obtain infinitely many maximal embedding dimension numerical semigroups, and that different numerical semigroups produce different maximal embedding dimension numerical semigroups. All maximal embedding dimension numerical semigroups are constructed in this way.

2 Arf numerical semigroups

A numerical semigroup S is *Arf* if for all $x,y,z \in S$, with $x \geq y \geq z$, $x + y - z$ is in S. In this section we present some characterizations of this property. For a numerical semigroup we will show how to compute the least Arf numerical semigroup containing it.

From Proposition 3.12 it follows that an Arf numerical semigroup has maximal embedding dimension.

Example 3.17. If m is a positive integer, then the numerical semigroup $\{0, m, \rightarrow\}$ is a numerical semigroup with the Arf property. Note that the semigroup T of Example 3.14 has maximal embedding dimension, while it is not Arf, because $14 + 14 - 13 = 15 \notin T$.

Given a positive integer x in a numerical semigroup S, the numerical semigroup $(x + S) \cup \{0\}$ is Arf if and only if S is Arf. This follows easily from the definition.

Proposition 3.18 ([63]). *Let S be a numerical semigroup and let $x \in S^*$. Then S is Arf if and only if $S' = (x + S) \cup \{0\}$ is Arf.*

In particular, S is Arf if and only if all the elements in $\mathscr{P}I(S)$ are Arf.

Let S be an Arf numerical semigroup. Then S has maximal embedding dimension. By Corollary 3.13 there exists a numerical semigroup S' and $x \in S' \setminus \{0\}$ such that $S = (x + S') \cup \{0\}$. If $S \neq \mathbb{N}$, then $S \subsetneq S'$. In view of Proposition 3.18, S' is also an Arf numerical semigroup. We can repeat this argument with S', and obtain an Arf numerical semigroup S'' and $y \in S'' \setminus \{0\}$ such that $S' = (y + S'') \cup \{0\}$. As $\mathbb{N} \setminus S$ has finitely many elements, this process is finite, obtaining in this way a stationary ascending chain of Arf numerical semigroups: $S_0 = S \subsetneq S_1 \subsetneq \cdots \subsetneq S_n = \mathbb{N}$, with $S_i = (x_{i+1} + S_{i+1}) \cup \{0\}$ for some $x_{i+1} \in S_{i+1} \setminus \{0\}$. The following statement can be derived from this idea.

Corollary 3.19. *Let S be a proper subset of $\mathbb{N}$. Then S is an Arf numerical semigroup if and only if there exist positive integers $x_1, \ldots, x_n$ such that*

$$S = \{0, x_1, x_1 + x_2, \ldots, x_1 + \cdots + x_{n-1}, x_1 + \cdots + x_n, \rightarrow\}$$

and $x_i \in \{x_{i+1}, x_{i+1} + x_{i+2}, \ldots, x_{i+1} + \cdots + x_n, \rightarrow\}$ for all $i \in \{1, \ldots, n\}$.

Proof. Necessity. Follows from the construction of the chain $S = S_0 \subsetneq S_1 \subsetneq \cdots \subsetneq S_n = \mathbb{N}$, with $S_i = (x_{i+1} + S_{i+1}) \cup \{0\}$ and $x_{i+1} \in S_{i+1} \setminus \{0\}$.

Sufficiency. Note that

$$S = (x_1 + (x_2 + (\cdots + ((x_n + \mathbb{N}) \cup \{0\}) \cdots)) \cup \{0\}.$$

As $\mathbb{N}$ is Arf, by applying Proposition 3.18 several times, we obtain that S is Arf. $\square$

Example 3.20. Take $x_1 = 7$, $x_2 = 4$ and $x_3 = 2$. This sequence fulfills the condition of Corollary 3.19. Then $S = \{0, 7, 11, 13, \rightarrow\}$ is a numerical semigroup with the Arf property. In view of Proposition 3.18, $(7 + S) \cup \{0\}, (11 + S) \cup \{0\}, (13 + S) \cup \{0\}, \ldots$ are Arf numerical semigroups as well. Proposition 3.18 also states that $T = -7 + S^*$ is an Arf numerical semigroup, because so is $S = (7 + T) \cup \{0\}$.

Recall (see Exercise 2.2) that the intersection of finitely many numerical semigroups is a numerical semigroup.

Example 3.21. It can be easily seen that

$$\bigcap_{n \in \mathbb{N}} \langle n, n+1 \rangle = \{0\}.$$

Hence the above result does not extend to the intersection of arbitrary families of numerical semigroups.

From the definition it also follows that the intersection of finitely many Arf numerical semigroups is again Arf.

Proposition 3.22. *The intersection of finitely many Arf numerical semigroups is an Arf numerical semigroup.*

Let S be a numerical semigroup. Since the complement of S in $\mathbb{N}$ is finite, the set of Arf numerical semigroups containing S is also finite. Proposition 3.22 ensures that the intersection of these semigroups is again an Arf numerical semigroup (it is actually one of them). We will denote this intersection by $\mathrm{Arf}(S)$ and we will refer to it as the *Arf closure* of S. Observe that the Arf closure of S is the smallest (with respect to set inclusion) Arf numerical semigroup containing S.

If X is a nonempty subset of nonnegative integers with $\gcd(X) = 1$, then $\langle X \rangle$ is a numerical semigroup. Any Arf numerical semigroup containing X must contain $\langle X \rangle$. So it makes sense to talk about the *Arf closure of* X, and define it as $\mathrm{Arf}(\langle X \rangle)$. We make an abuse of notation and will write $\mathrm{Arf}(X)$ to denote $\mathrm{Arf}(\langle X \rangle)$.

Computing the set of numerical semigroups that contain a given numerical semigroup can be tedious. Even more if one has to decide which are Arf among them, and then either compute the intersection of them all or decide which is the smallest. We now describe an alternative way introduced in [88] to compute the Arf closure that is much more efficient.

Lemma 3.23. *Let S be a submonoid of $\mathbb{N}$. Then*

$$S' = \{x + y - z \mid x, y, z \in S, x \geq y \geq z\}$$

is a submonoid of $\mathbb{N}$ and $S \subseteq S'$.

Proof. Let $x \in S$. Then $x + x - x \in S'$, whence $S \subseteq S'$. Clearly $S' \subseteq \mathbb{N}$. Now take $a, b \in S'$. By the definition of S', there exist $x_1, x_2, y_1, y_2, z_1, z_2 \in S$, such that $x_i \geq y_i \geq z_i$, $i \in \{1, 2\}$, and $a = x_1 + y_1 - z_1$, $b = x_2 + y_2 - z_2$. Hence, $a + b = (x_1 + x_2) + (y_1 + y_2) - (z_1 + z_2)$. Clearly $x_1 + x_2, y_1 + y_2, z_1 + z_2 \in S$ and $x_1 + x_2 \geq y_1 + y_2 > z_1 + z_2$. This proves that $a + b \in S'$. $\qquad\square$

For a given submonoid S of $\mathbb{N}$ and $n \in \mathbb{N}$, define S^n recurrently as follows:

- $S^0 = S$,
- $S^{n+1} = (S^n)'$.

We see that this becomes stationary at the Arf closure of S.

Proposition 3.24. *Let S be a numerical semigroup. Then there exists $k \in \mathbb{N}$ such that $S^k = \mathrm{Arf}(S)$.*

Proof. By using induction on n, it can be easily proved that $S^n \subseteq \mathrm{Arf}(S)$ for all $n \in \mathbb{N}$. By Lemma 3.23, $S^n \subseteq S^{n+1}$ and $S \subseteq S^n$ for all $n \in \mathbb{N}$. As we pointed out before, the number of numerical semigroups containing S is finite, whence $S^k = S^{k+1}$ for some $k \in \mathbb{N}$. It follows that S^k is an Arf numerical semigroup. As $S^k \subseteq \mathrm{Arf}(S)$ and $\mathrm{Arf}(S)$ is the smallest Arf numerical semigroup containing S, we conclude that $S^k = \mathrm{Arf}(S)$. $\qquad\square$

Although this is a nice characterization, we have not yet shown how to compute S^k. So more effort is needed to find an effective way to compute the Arf closure of a numerical semigroup.

Lemma 3.25. *Let* $m, r_1, \ldots, r_p, n \in \mathbb{N}$ *such that* $\gcd(\{m, r_1, \ldots, r_p\}) = 1$. *Then*

$$m + \langle m, r_1, \ldots, r_p \rangle^n \subseteq \mathrm{Arf}(m, m + r_1, \ldots, m + r_p).$$

Proof. We use once more induction on n. For $n = 0$ we have to prove that $m + \langle m, r_1, \ldots, r_p \rangle \subseteq \mathrm{Arf}(m, m + r_1, \ldots, m + r_p)$. Let $i, j \in \{1, \ldots, p\}$. Then $m, m + r_i, m + r_j \in \mathrm{Arf}(m, m + r_1, \ldots, m + r_p)$, whence $m + r_i + r_j = (m + r_i) + (m + r_j) - m \in \mathrm{Arf}(m, m + r_1, \ldots, m + r_p)$. Now, for $k \in \{1, \ldots, p\}$, $m, m + r_i + r_j, m + r_k \in \mathrm{Arf}(m, m + r_1, \ldots, m + r_k)$, and therefore $m + r_i + r_j + r_k = (m + r_i + r_j) + (m + r_k) - m \in \mathrm{Arf}(m, m + r_1, \ldots, m + r_k)$. By repeating this argument we obtain that for every $a, a_1, \ldots, a_p \in \mathbb{N}$, we have that $(a + 1)m + a_1 r_1 + \cdots + a_p r_p \in \mathrm{Arf}(m, m + r_1, \ldots, m + r_p)$, and thus $m + \langle m, r_1, \ldots, r_p \rangle \subseteq \mathrm{Arf}(m, m + r_1, \ldots, m + r_p)$.

Now assume that $m + \langle m, r_1, \ldots, r_p \rangle^n \subseteq \mathrm{Arf}(m, m + r_1, \ldots, m + r_p)$ and let us prove that $m + \langle m, r_1, \ldots, r_p \rangle^{n+1} \subseteq \mathrm{Arf}(m, m + r_1, \ldots, m + r_p)$. Let $a \in m + \langle m, r_1, \ldots, r_p \rangle^{n+1}$. Then $a = m + b$ with $b \in \langle m, r_1, \ldots, r_p \rangle^{n+1}$. Hence there exist $x, y, z \in \langle m, r_1, \ldots, r_p \rangle^n$ such that $x \geq y \geq z$ and $x + y - z = b$. In this way $a = m + b = m + x + y - z = (m + x) + (m + y) - (m + z) \in \mathrm{Arf}(m, m + r_1, \ldots, m + r_p)$, since by induction hypothesis $m + x, m + y, m + z \in m + \langle m, r_1, \ldots, r_p \rangle^n \subseteq \mathrm{Arf}(m, m + r_1, \ldots, m + r_p)$. $\qquad\square$

From this we can give a procedure to compute Arf closures that has an extended Euclid's algorithm taste.

Proposition 3.26. *Let* $m, r_1, \ldots, r_p$ *be nonnegative integers with greatest common divisor one. Then*

$$\mathrm{Arf}(m, m + r_1, \ldots, m + r_p) = (m + \mathrm{Arf}(m, r_1, \ldots, r_p)) \cup \{0\}.$$

Proof. By using Proposition 3.24 and Lemma 3.25, we obtain that $(m + \mathrm{Arf}(m, r_1, \ldots, r_p)) \cup \{0\} \subseteq \mathrm{Arf}(m, m + r_1, \ldots, m + r_p)$. For the other inclusion observe that $m, m + r_1, \ldots, m + r_p \in (m + \mathrm{Arf}(m, r_1, \ldots, r_p)) \cup \{0\}$, which by Proposition 3.18 is an Arf numerical semigroup. It follows that $\mathrm{Arf}(m, m + r_1, \ldots, m + r_p) \subseteq (m + \mathrm{Arf}(m, r_1, \ldots, r_p)) \cup \{0\}$. $\qquad\square$

The Frobenius number of the Arf closure can then be computed as follows.

Corollary 3.27. *Let* $m, r_1, \ldots, r_p$ *be nonnegative integers with greatest common divisor one. Then*

$$\mathrm{F}(\mathrm{Arf}(m, m+r_1, \ldots, m+r_p)) = m + \mathrm{F}(\mathrm{Arf}(m, r_1, \ldots, r_p)).$$

We have now all the ingredients needed to give a recursive way of calculating the elements of the Arf closure of any subset of nonnegative integers with greatest common divisor one. Let $X \subseteq \mathbb{N} \setminus \{0\}$ be such that $\gcd(X) = 1$. Define recursively the following sequence of subsets of $\mathbb{N}$:

- $A_1 = X$,
- $A_{n+1} = (\{x - \min A_n \mid x \in A_n\} \setminus \{0\}) \cup \{\min A_n\}$.

As a consequence of Euclid's algorithm for the computation of $\gcd(X)$, we obtain that there exists $q = \min\{k \in \mathbb{N} \mid 1 \in A_k\}$.

Proposition 3.28. *With the above notation, we have that*

$$\mathrm{Arf}(X) = \{0, \min A_1, \min A_1 + \min A_2, \ldots, \min A_1 + \cdots + \min A_{q-1}, \rightarrow\}.$$

Proof. Since $1 \in A_q$, $\mathrm{Arf}(A_q) = \mathbb{N}$. Hence by Proposition 3.26, $\mathrm{Arf}(A_{q-1}) = (\min A_{q-1} + \mathbb{N}) \cup \{0\}$. This implies

$$\mathrm{Arf}(A_{q-1}) = \{0, \min A_{q-1}, \rightarrow\}.$$

Assume as induction hypothesis that

$$\mathrm{Arf}(A_{q-i}) = \{0, \min A_{q-i}, \min A_{q-i} + \min A_{q-i+1}, \ldots,$$
$$\min A_{q-i} + \cdots + \min A_{q-1}, \rightarrow\}.$$

We must prove now that

$$\mathrm{Arf}(A_{q-i-1}) = \{0, \min A_{q-i-1}, \min A_{q-i-1} + \min A_{q-i}, \ldots,$$
$$\min A_{q-i-1} + \cdots + \min A_{q-1}, \rightarrow\}.$$

By Proposition 3.26, we know that $\mathrm{Arf}(A_{q-i-1}) = (\min A_{q-i-1} + \mathrm{Arf}(A_{q-i})) \cup \{0\}$. By using now the induction hypothesis and Corollary 3.27, we obtain the desired result. $\qquad\square$

Example 3.29 ([88]). Let us compute $\mathrm{Arf}(7, 24, 33)$.

$A_1 = \{7, 24, 33\}, \ \min_{\leq} A_1 = 7,$
$A_2 = \{7, 17, 26\}, \ \min_{\leq} A_2 = 7,$
$A_3 = \{7, 10, 19\}, \ \min_{\leq} A_3 = 7,$
$A_4 = \{7, 3, 12\}, \ \min_{\leq} A_4 = 3,$
$A_5 = \{4, 3, 9\}, \ \min_{\leq} A_5 = 3,$
$A_6 = \{1, 3, 6\},$

whence $\mathrm{Arf}(7, 24, 33) = \{0, 7, 14, 21, 24, 27, \rightarrow\}$.

3 Saturated numerical semigroups

A numerical semigroup S is *saturated* if the following condition holds: if $s, s_1, \ldots, s_r \in S$ are such that $s_i \leq s$ for all $i \in \{1, \ldots, r\}$ and $z_1, \ldots, z_r \in \mathbb{Z}$ are such that $z_1 s_1 + \cdots + z_r s_r \geq 0$, then $s + z_1 s_1 + \cdots + z_r s_r \in S$.

Example 3.30. The semigroup $S = \langle 7, 11, 13, 15, 16, 17, 19 \rangle$ appearing in Example 3.20 is an Arf semigroup but it is not saturated. Note that $7, 11 \in S$ and $12 = 11 + 2 \times 11 - 3 \times 7 \notin S$.

From the definition it easily follows that every saturated numerical semigroup is Arf, and thus it has maximal embedding dimension.

Lemma 3.31. *Every saturated numerical semigroup has the Arf property.*

Next we describe a characterization of this kind of semigroup that appears in [89].

For $A \subseteq \mathbb{N}$ and $a \in A \setminus \{0\}$, set

$$d_A(a) = \gcd\{x \in A \mid x \leq a\}.$$

Lemma 3.32. *Let S be a saturated numerical semigroup and let $s \in S$. Then $s + d_S(s) \in S$.*

Proof. Let $\{s_1, \ldots, s_r\} = \{x \in S \mid x \leq s\}$. By Bézout's identity, there exist integers $z_1, \ldots, z_r$ such that $z_1 s_1 + \cdots + z_r s_r = d_S(s)$. As S is saturated, we get $s + d_S(s) \in S$. □

We are going to see that this property characterizes saturated numerical semigroups. First we need some previous lemmas.

Lemma 3.33. *Let A be a nonempty subset of positive integers such that $\gcd(A) = 1$ and $a + d_A(a) \in A$ for all $a \in A$. Then $a + k d_A(a) \in A$ for all $k \in \mathbb{N}$, and $A \cup \{0\}$ is a numerical semigroup.*

Proof. We proceed by induction on $d_A(a)$.

Note that $d_A(a) > 0$. We show that if $d_A(a) = 1$, then $a + k \in A$ for all $k \in \mathbb{N}$. To this end we use induction on k. For $k = 0$, the result is clear. Assume that $a + k \in A$. Since $0 \neq d_A(a+k) \leq d_A(a) = 1$, we have that $d_A(a+k) = 1$. Hence $a + k + 1 = a + k + d_A(a+k) \in A$.

Now assume that if $a' \in A$ and $d_A(a') < d_A(a)$, then $a' + k d_A(a') \in A$ for all $k \in \mathbb{N}$. Thus, suppose that $d_A(a) \geq 2$ and let us prove that $a + k d_A(a) \in A$ for all $k \in \mathbb{N}$. Since $\gcd(A) = 1$, there exists $b \in A$ such that $d_A(b) = 1$. If $d_A(a + k d_A(a)) = d_A(a)$ and $a + k d_A(a) \in A$, then $a + (k+1) d_A(a) = a + k d_A(a) + d_A(a + k d_A(a)) \in A$. From these two remarks, we deduce that there exists a least positive integer t such that $a + t d_A(a) \in A$ and $d_A(a + t d_A(a)) < d_A(a)$. As $d_A(a + t d_A(a)) < d_A(a)$, by applying the induction hypothesis, we obtain that $(a + t d_A(a)) + k d_A(a + t d_A(a)) \in A$ for all $k \in \mathbb{N}$. Clearly, $d_A(a + t d_A(a))$ divides $d_A(a)$, whence $d_A(a) = l d_A(a + t d_A(a))$ for some positive integer l. Consequently, $a + t\mathrm{d}_A(a) + \frac{k}{l}\mathrm{d}_A(a) \in A$ for all $k \in \mathbb{N}$, and thus $a + (t + n)\mathrm{d}_A(a) \in A$ for all $n \in \mathbb{N}$. From the definition of t, it follows that $a + k\mathrm{d}_A(a) \in A$ for all $k \in \{0,\dots,t\}$. We conclude that $a + k\mathrm{d}_A(a) \in A$ for all $k \in \mathbb{N}$.

Finally, let us prove that $A \cup \{0\}$ is a numerical semigroup. Since $\gcd(A) = 1$, it suffices to prove that for any $a, b \in A$, $a + b \in A$. Assume that $a \leq b$. Then $\mathrm{d}_A(b)$ divides $\mathrm{d}_A(a)$ and thus there exists $\lambda \in \mathbb{N}$ such that $\mathrm{d}_A(a) = \lambda \mathrm{d}_A(b)$. Note also that $\mathrm{d}_A(a)$ divides a, whence $a = \mu \mathrm{d}_A(a)$ for some $\mu \in \mathbb{N}$. Therefore $a + b = \mu \mathrm{d}_A(a) + b = \mu \lambda \mathrm{d}_A(b) + b$, which, as we have just proven, belongs to A. $\qquad\square$

Proposition 3.34. *Let A be a nonempty subset of $\mathbb{N}$ such that $0 \in A$ and $\gcd(A) = 1$. The following conditions are equivalent.*

1) A is a saturated numerical semigroup.
2) $a + \mathrm{d}_A(a) \in A$ for all $a \in A \setminus \{0\}$.
3) $a + k\mathrm{d}_A(a) \in A$ for all $a \in A \setminus \{0\}$ and $k \in \mathbb{N}$.

Proof. 1) implies 2). Follows from Lemma 3.32.

2) implies 3). Follows from Lemma 3.33.

3) implies 1). By Lemma 3.33, we know that A is a numerical semigroup. Let $a, a_1, \dots, a_r \in A$ with $a_i \leq a$ for all $i \in \{1, \dots, r\}$, and let $z_1, \dots, z_r$ be integers such that $z_1 a_1 + \cdots + a_r z_r \geq 0$. Since $a_i \leq a$, it follows that $\mathrm{d}_A(a)$ divides a_i for all $i \in \{1, \dots, r\}$. Hence, there exists $k \in \mathbb{N}$ such that $z_1 a_1 + \cdots + z_r a_r = k\mathrm{d}_A(a)$, and thus $a + z_1 a_1 + \cdots + z_r a_r = a + k\mathrm{d}_A(a) \in A$. This proves that A is saturated. $\qquad\square$

We now focus on obtaining a similar characterization as the one given in Proposition 3.18 and Corollary 3.19 for Arf numerical semigroups. As we will see in this setting the characterization is not so generous.

Proposition 3.35 ([63]). *Let S be a numerical semigroup. The following conditions are equivalent.*

1) S is saturated.
2) There exists $x \in S^*$ such that $(x + S) \cup \{0\}$ is a saturated numerical semigroup.

Proof. 1) implies 2). Assume that $S = \{0 < s_1 < s_2 < \cdots < s_n < \cdots\}$. We prove that $(s_1 + S) \cup \{0\} = \{0 < s_1 < s_1 + s_1 < s_1 + s_2 < \cdots < s_1 + s_n < \cdots\}$ is saturated. In view of Proposition 3.34, it suffices to show that for all $n \in \mathbb{N}$, the element $s_1 + s_n + \gcd\{0, s_1, s_1 + s_1, \dots, s_1 + s_n\}$ lies in $(s_1 + S) \cup \{0\}$. Since S is saturated, $s_n + \gcd\{0, s_1, \dots, s_n\} \in S$. Moreover $\gcd\{0, s_1, s_1 + s_1, s_1 + s_2, \dots, s_1 + s_n\} = \gcd\{0, s_1, s_2, \dots, s_n\}$, whence $s_1 + s_n + \gcd\{0, s_1, s_1 + s_1, \dots, s_1 + s_n\} \in (s_1 + S) \cup \{0\}$.

2) implies 1). If $S = \{0 < s_1 < \cdots < s_n < \cdots\}$, then $(x + S) \cup \{0\} = \{0 < x < s_1 + x < \cdots < s_n + x < \cdots\}$. Since $\gcd\{0, x, x + s_1, \dots, x + s_n\} = \gcd\{0, x, s_1, \dots, s_n\}$, we have that $\gcd\{0, x, x + s_1, \dots, x + s_n\}$ divides $\gcd\{0, s_1, \dots, s_n\}$, namely, there exists $k \in \mathbb{N}$ such that $k(\gcd\{0, x, x + s_1, \dots, x + s_n\}) = \gcd\{0, s_1, \dots, s_n\}$. By Proposition 3.34, if we want to prove that S is saturated, it suffices to show that $s_n + \gcd\{0, s_1, \dots, s_n\} \in S$ for all n. As $(x + S) \cup \{0\}$ is saturated, by Proposition 3.34, we have that $x + s_n + k(\gcd\{0, x, x + s_1, \dots, x + s_n\}) \in (x + S) \cup \{0\}$ and thus $s_n + \gcd\{0, s_1, \dots, s_n\} \in S$. $\qquad\square$

From the proof of this result we obtain the following consequence.

Corollary 3.36. *Let S be a numerical semigroup. Then S is saturated if and only if* $(m(S) + S) \cup \{0\}$ *is saturated.*

Example 3.37. The semigroup $S = \langle 5, 7, 8, 9, 11 \rangle$ is a saturated numerical semigroup. From Corollary 3.36 we have that both $(5 + S) \cup \{0\}$ and $-5 + S^*$ are saturated.

Corollary 3.38. *Let S be a proper subset of* $\mathbb{N}$. *Then S is a saturated numerical semigroup if and only if there exist positive integers* $x_1, \ldots, x_n$ *such that*

$$S = \{0, x_1, x_1 + x_2, \ldots, x_1 + \cdots + x_n, \rightarrow\}$$

and

$$\gcd\{x_1, \ldots, x_k\} \in \{x_{k+1}, x_{k+1} + x_{k+2}, \ldots, x_{k+1} + \cdots + x_n, \rightarrow\}$$

for all $k \in \{1, \ldots, n\}$.

Proof. Necessity. Since S is a saturated numerical semigroup, S is also Arf, whence by Corollary 3.19 there exist positive integers $x_1, \ldots, x_n$ such that

$$S = \{0, x_1, x_1 + x_2, \ldots, x_1 + \cdots + x_n, \rightarrow\}.$$

As S is saturated, for all $k \in \{1, \ldots, n\}$, $(x_1 + \cdots + x_k) + \gcd\{0, x_1, x_1 + x_2, \ldots, x_1 + \cdots + x_k\} \in S$ and since $\gcd\{0, x_1, x_1 + x_2, \ldots, x_1 + \cdots + x_k\} = \gcd\{x_1, \ldots, x_k\}$, we have that $(x_1 + \cdots + x_k) + \gcd\{x_1, \ldots, x_k\} \in \{0, x_1, x_1 + x_2, \ldots, x_1 + \cdots + x_n, \rightarrow\}$, or equivalently, $\gcd\{x_1, \ldots, x_k\} \in \{x_{k+1}, x_{k+1} + x_{k+2}, \ldots, x_{k+1} + \cdots + x_n, \rightarrow\}$.

Sufficiency. By using Proposition 3.34, it suffices to show that $(x_1 + \cdots + x_k) + \gcd\{0, x_1, x_1 + x_2, \ldots, x_1 + \cdots + x_k\} \in S$ for all $k \in \{1, \ldots, n\}$. As pointed out above, this is equivalent to prove that $(x_1 + \cdots + x_k) + \gcd(x_1, \ldots, x_k) \in S$, and this follows from the hypothesis.   □

As for Arf numerical semigroups, the intersection of finitely many saturated numerical semigroups is again saturated. This follows easily from the definition.

Proposition 3.39. *The intersection of finitely many saturated numerical semigroups is a saturated numerical semigroup.*

This allows us to define the saturated closure of a numerical semigroup (or of a subset of nonnegative integers with greatest common divisor one), as we did for Arf numerical semigroups. Given a numerical semigroup S, we denote by $\mathrm{Sat}(S)$ the intersection of all saturated numerical semigroups containing S, or in other words, the smallest (with respect to set inclusion) saturated numerical semigroup containing S. We call this semigroup the *saturated closure* of S.

The saturated closure of a semigroup (or of any set of nonnegative integers with greatest common divisor one) can be computed as follows.

Proposition 3.40. *Let $n_1 < n_2 < \cdots < n_e$ be positive integers such that $\gcd(n_1,\ldots,n_e) = 1$. For every $i \in \{1,\ldots,e\}$, set $d_i = \gcd(n_1,\ldots,n_i)$ and for all $j \in \{1,\ldots,p-1\}$ define $k_j = \max\{k \in \mathbb{N} \mid n_j + kd_j < n_{j+1}\}$. Then*

$$\mathrm{Sat}(n_1,\ldots,n_e) = \{0, n_1, n_1 + d_1, \ldots, n_1 + k_1 d_1, n_2, n_2 + d_2, \ldots, n_2 + k_2 d_2,$$
$$\ldots, n_{e-1}, n_{e-1} + d_{e-1}, \ldots, n_{e-1} + k_{e-1} d_{e-1}, n_e, n_e + 1, \rightarrow\}.$$

Proof. Let

$$A = \{0, n_1, n_1 + d_1, \ldots, n_1 + k_1 d_1, n_2, n_2 + d_2, \ldots, n_2 + k_2 d_2,$$
$$\ldots, n_{e-1}, n_{e-1} + d_{e-1}, \ldots, n_{p-1} + k_{e-1} d_{e-1}, n_e, n_e + 1, \rightarrow\}.$$

Clearly A is not empty, $0 \in A$, $\gcd(A) = 1$ and $a + d_A(a) \in A$ for all $a \in A$. By Proposition 3.34, A is a saturated numerical semigroup, and as $\{n_1,\ldots,n_e\} \subset A$, we get that $\mathrm{Sat}(n_1,\ldots,n_e) \subseteq A$. For the other inclusion, take $a \in A$. Then there exists $i \in \{1,\ldots,e\}$ and $k \in \mathbb{N}$ such that $a = n_i + kd_i$ (note that $d_e = 1$). Since $\{n_1,\ldots,n_e\} \subset \mathrm{Sat}(n_1,\ldots,n_e)$, we have that $d_{\mathrm{Sat}(n_1,\ldots,n_e)}(n_i)$ divides d_i, whence there exists $l \in \mathbb{N}$ such that $d_i = l d_{\mathrm{Sat}(n_1,\ldots,n_e)}(n_i)$. From Proposition 3.34, we know that $n_i + t d_{\mathrm{Sat}(n_1,\ldots,n_e)}(n_i) \in \mathrm{Sat}(n_1,\ldots,n_e)$ for all $t \in \mathbb{N}$ and thus $a = n_i + kd_i = n_i + kl d_{\mathrm{Sat}(n_1,\ldots,n_e)}(n_i) \in \mathrm{Sat}(n_1,\ldots,n_e)$. $\quad\square$

Example 3.41. $\mathrm{Sat}(\{12,20,26,35\}) = \{0,12,20,24,26,28,30,32,34,35,\rightarrow\}$.

Exercises

Exercise 3.1 ([45, 87]). Let m be an integer greater than or equal to two, and let $(k_1,\ldots,k_{m-1}) \in \mathbb{N}^{m-1}$. Prove that $\{0, k_1 m + 1, \ldots, k_{m-1} m + m - 1\}$ is the Apéry set of m in a numerical semigroup with multiplicity m if and only if $(k_1,\ldots,k_{m-1})$ is a solution to the system of inequalities

$$
\begin{array}{ll}
x_i \geq 1 & \text{for all } i \in \{1,\ldots,m-1\}, \\
x_i + x_j - x_{i+j} \geq 0 & \text{for all } i,j \text{ with } 1 \leq i \leq j \leq m-1, i+j \leq m-1, \\
x_i + x_j - x_{i+j-m} \geq -1 & \text{for all } i,j \text{ with } 1 \leq i \leq j \leq m-1, i+j > m.
\end{array}
$$

Exercise 3.2 ([87]). Let m be an integer greater than or equal to two, and let $(k_1,\ldots,k_{m-1}) \in \mathbb{N}^{m-1}$. Show that $S = \langle m, k_1 m + 1, \ldots, k_{m-1} m + m - 1 \rangle$ is a numerical semigroup with multiplicity m and maximal embedding dimension if and only if $(k_1,\ldots,k_{m-1})$ is a solution to the system of inequalities

$$
\begin{array}{ll}
x_i \geq 1 & \text{for all } i \in \{1,\ldots,m-1\}, \\
x_i + x_j - x_{i+j} \geq 1 & \text{for all } i,j \text{ with } 1 \leq i \leq j \leq m-1, i+j \leq m-1, \\
x_i + x_j - x_{i+j-m} \geq 0 & \text{for all } i,j \text{ with } 1 \leq i \leq j \leq m-1, i+j > m.
\end{array}
$$

Exercise 3.3. Let m be a positive integer. Prove that the intersection of finitely many maximal embedding dimension numerical semigroups with multiplicity m is again a maximal embedding dimension numerical semigroup with multiplicity m.

Show with an example that the intersection of finitely many maximal embedding dimension numerical semigroups might not have maximal embedding dimension.

Exercise 3.4. Let S be a numerical semigroup and let m be a nonzero element of S. Denote by $Ap_0(S,m) = \{w \in Ap(S,m) \mid w \text{ is even}\}$ and by $Ap_1(S,m) = \{w \in Ap(S,m) \mid w \text{ is odd}\}$. Prove that

a) if m is even, then the cardinalities of $Ap_0(S,m)$ and $Ap_1(S,m)$ are the same and equal to $\frac{m}{2}$,

b) if m is odd and $Ap_1(S,m) = m - 1$, then S is a maximal embedding dimension of multiplicity m.

Exercise 3.5. Let S be a numerical semigroup. Show that if S has maximal embedding dimension, has the Arf property or it is saturated, then so is $S \cup \{F(S)\}$.

Exercise 3.6. Let S be numerical semigroup and let $m \in S \setminus \{0,1\}$. Set $T = (m + S) \cup \{0\}$ (see Exercise 2.9). Prove that

a) $F(T) = F(S) + m$,

b) $g(T) = g(S) + m - 1$.

Exercise 3.7. Let S be a numerical semigroup with maximal ideal M (see Exercise 2.13) and multiplicity m. Prove that S has maximal embedding dimension if and only if $M - M = -m + M$. Show that this is also equivalent to $m + M = M + M (= \{x + y \mid x, y \in M\})$.

Exercise 3.8. Let S be a maximal embedding dimension numerical semigroup with minimal system of generators $\{n_1 < n_2 < \cdots < n_e\}$. Prove that $g(S) \geq \frac{n_e - 1}{2}$.

Exercise 3.9. Let $S = \langle 5, 7, 9 \rangle$. Compute the smallest (with respect to set inclusion) maximal embedding dimension numerical semigroup with multiplicity 5 containing S.

Exercise 3.10. Prove that $\{\langle 2, 2k+1 \rangle \mid k \in \mathbb{N}\}$ is the set of all maximal embedding dimension numerical semigroups of type 1. Which is the set of all Arf numerical semigroups of type 1? And that of saturated numerical semigroups?

Exercise 3.11. Check that $\langle 4, 9, 10, 11 \rangle$ is the smallest (with respect to set inclusion) Arf numerical semigroup containing $\langle 4, 9, 11 \rangle$.

Exercise 3.12. Show that $\langle 10, 14, 16, 18, 22, 27, 29, 31, 33, 35 \rangle$ is the smallest (with respect to set inclusion) saturated numerical semigroup containing $\langle 10, 14, 27 \rangle$.

Exercise 3.13 ([8]). Prove that every Arf numerical semigroup that is not a half-line is acute (see Exercise 2.8).

Chapter 3
Irreducible numerical semigroups

Introduction

Symmetric numerical semigroups are probably the numerical semigroups that have been most studied in the literature. The motivation and introduction of these semigroups is due mainly to Kunz, who in his manuscript [44] proves that a one-dimensional analytically irreducible Noetherian local ring is Gorenstein if and only if its value semigroup is symmetric. Symmetric numerical semigroups always have odd Frobenius number. The translation of this concept for numerical semigroups with even Frobenius number motivates the definition of pseudo-symmetric numerical semigroups. In [5] it is shown that these semigroups also have their interpretation in one-dimensional local rings, since a numerical semigroup is pseudo-symmetric if and only if its semigroup ring is a Kunz ring.

Irreducible numerical semigroups gather both symmetric and pseudo-symmetric numerical semigroups. This concept was introduced in [73]. Its study is clearly well motivated from the semigroup theory point of view as the reader will see from the definition.

1 Symmetric and pseudo-symmetric numerical semigroups

A numerical semigroup is *irreducible* if it cannot be expressed as the intersection of two numerical semigroups properly containing it.

We are going to show that irreducible numerical semigroups are maximal in the set of numerical semigroups with fixed Frobenius number. First we prove that adding the Frobenius number to a numerical semigroup yields a numerical semigroup. This is a particular case of a more general result that we will present later.

Lemma 4.1. *Let S be a numerical semigroup other than* $\mathbb{N}$. *Then* $S \cup \{F(S)\}$ *is again a numerical semigroup.*

Proof. The complement of $S \cup \{F(S)\}$ in $\mathbb{N}$ is finite, because $\mathbb{N} \setminus S$ is finite. Take $a, b \in S \cup \{F(S)\}$. If any of them is $F(S)$, then $a + b \geq F(S)$ and thus $a + b \in S \cup \{F(S)\}$. If both a and b are in S, then $a + b \in S \subseteq S \cup \{F(S)\}$. As $0 \in S \cup \{F(S)\}$, this proves that $S \cup \{F(S)\}$ is a numerical semigroup. $\quad\square$

Theorem 4.2 ([73]). *Let S be a numerical semigroup. The following conditions are equivalent.*

1) S is irreducible.
2) S is maximal in the set of all numerical semigroups with Frobenius number $F(S)$.
3) S is maximal in the set of all numerical semigroups that do not contain $F(S)$.

Proof. *1) implies 2).* Let T be a numerical semigroup such that $S \subseteq T$ and $F(T) = F(S)$. Then $S = (S \cup \{F(S)\}) \cap T$. Since S is irreducible, we deduce that $S = T$.

2) implies 3). Let T be a numerical semigroup fulfilling that $S \subseteq T$ and $F(S) \notin T$. Then $T \cup \{F(S) + 1, F(S) + 2, \rightarrow\}$ is a numerical semigroup that contains S with Frobenius number $F(S)$. Therefore, $S = T \cup \{F(S) + 1, F(S) + 2, \rightarrow\}$ and so $S = T$.

3) implies 1). Let S_1 and S_2 be two numerical semigroups that contain S properly. Then, by hypothesis, $F(S) \in S_1$ and $F(S) \in S_2$. Hence $S \neq S_1 \cap S_2$. $\quad\square$

This result is also stated in [32, Proposition 4], but using a different terminology. We now introduce this terminology and see why both results are equivalent.

A numerical semigroup S is *symmetric* if it is irreducible and $F(S)$ is odd. We say that S is *pseudo-symmetric* provided that S is irreducible and $F(S)$ is even.

Given a numerical semigroup S, if S is not irreducible, then by Theorem 4.2, there exists an irreducible numerical semigroup T containing S with $F(S) = F(T)$. The following result can be viewed as a procedure to construct this irreducible numerical semigroup.

Lemma 4.3 ([58]). *Let S be a numerical semigroup and assume that there exists*

$$h = \max\{x \in \mathbb{Z} \setminus S \mid F(S) - x \notin S \text{ and } x \neq F(S)/2\}.$$

Then $S \cup \{h\}$ is a numerical semigroup with Frobenius number $F(S)$.

Proof. Clearly $S \cup \{h\}$ has finite complement in $\mathbb{N}$, and $0 \in S \cup \{h\}$. Let

$$H = \{x \in \mathbb{Z} \setminus S \mid F(S) - x \notin S \text{ and } x \neq F(S)/2\}.$$

If $x \in H$, then $F(S) - x \in H$. From this we deduce that $h > F(S)/2$.

Take $s \in S \setminus \{0\}$. If $h + s \notin S$, from the maximality of h, $F(S) - (h + s) = t \in S$ (as $h > F(S)/2, h + s \neq F(S)/2$). Hence $F(S) - h = t + s \in S$, contradicting the definition of h.

If $2h \notin S$, then again by the maximality of h, we get that $F(S) - 2h = t \in S$. As we have seen above, $h + t \in S$. However, $h + t = F(S) - h$, which cannot belong to S. This is a contradiction. $\quad\square$

The next proposition gives characterizations for the concepts of symmetric and pseudo-symmetric numerical semigroups. Sometimes in the literature these are chosen as the definitions.

Proposition 4.4. *Let S be a numerical semigroup.*

1) S is symmetric if and only if $F(S)$ *is odd and* $x \in \mathbb{Z} \setminus S$ *implies* $F(S) - x \in S$.
2) S is pseudo-symmetric if and only if $F(S)$ *is even and* $x \in \mathbb{Z} \setminus S$ *implies that either* $F(S) - x \in S$ *or* $x = F(S)/2$.

Proof. We only prove the first statement, because the second follows analogously.

Necessity. If there exists $x \in \mathbb{Z} \setminus S$ such that $F(S) - x \notin S$, then there exists the maximum h defined in Lemma 4.3. Hence $S \cup \{h\}$ is a numerical semigroup with Frobenius number $F(S)$, contradicting the maximality of S in Theorem 4.2.

Sufficiency. In view of Theorem 4.2, it suffices to prove that S is maximal in the set of all numerical semigroups that do not contain $F(S)$. Let T be a numerical semigroup with $S \subsetneq T$. Take $x \in T \setminus S \subset \mathbb{Z} \setminus S$. Then by hypothesis $F(S) - x \in S$ and thus $F(S) - x \in T$. But this implies that $F(S) = x + (F(S) - x) \in T$. □

From this result one easily deduces the following alternative characterization.

Corollary 4.5. *Let S be a numerical semigroup.*

1) S is symmetric if and only if $g(S) = \frac{F(S)+1}{2}$.
2) S is pseudo-symmetric if and only if $g(S) = \frac{F(S)+2}{2}$.

Remark 4.6. We know (Lemma 2.14) that if S is a numerical semigroup, then $g(S) \geq \frac{F(S)+1}{2}$. As a consequence of this corollary, we have that irreducible numerical semigroups are those numerical semigroups with the least possible genus in terms of their Frobenius number.

From Proposition 2.13 and Corollary 4.5, we obtain the following consequence.

Corollary 4.7. *Every numerical semigroup of embedding dimension two is symmetric.*

Example 4.8. The numerical semigroup $\langle 4, 6, 7 \rangle = \{0, 4, 6, 7, 8, 10, 11, \rightarrow\}$ is symmetric, $\langle 3, 4, 5 \rangle = \{0, 3, \rightarrow\}$ is pseudo-symmetric, and $\langle 5, 7, 9 \rangle$ is not irreducible.

The Apéry sets of irreducible numerical semigroups have special shapes. This shape characterizes them as we see in the rest of this section.

Lemma 4.9. *Let S be a numerical semigroup and let n be a positive integer of S. If* $x, y \in S$ *are such that* $x + y \in \mathrm{Ap}(S, n)$, *then* $\{x, y\} \subseteq \mathrm{Ap}(S, n)$.

Proof. This is an immediate consequence of the definition of Apéry set. □

Proposition 4.10. *Let S be a numerical semigroup and let n be a positive integer of S. Let* $\mathrm{Ap}(S, n) = \{a_0 < a_1 < \cdots < a_{n-1}\}$ *be the Apéry set of n in S. Then S is symmetric if and only if* $a_i + a_{n-1-i} = a_{n-1}$ *for all* $i \in \{0, \ldots, n-1\}$.

Proof. Necessity. By Proposition 2.12, we know that $F(S) = a_{n-1} - n$. As $a_i - n \notin S$ and S is symmetric, $F(S) - (a_i - n) = a_{n-1} - a_i \in S$. By Lemma 4.9, we deduce that there exists $j \in \{0, \ldots, n-1\}$ such that $a_{n-1} = a_i + a_j$. Since $a_0 < a_1 < \cdots < a_{n-1}$, j must be $n - 1 - i$.

Sufficiency. From the hypothesis, we deduce that $\{a_{n-1}\} = \text{Maximals}_{\leq_S} \text{Ap}(S, n)$. By Proposition 2.20, $PF(S) = \{F(S)\}$, and thus $\{F(S)\} = \text{Maximals}_{\leq_S}(\mathbb{Z} \setminus S)$. This in particular implies that if $x \in \mathbb{Z} \setminus S$, then $F(S) - x \in S$. Besides, if $F(S)/2$ is an integer, then $F(S)/2 \in \mathbb{Z} \setminus S$. We have just shown that this would imply that $F(S) - F(S)/2 = F(S)/2 \in S$, a contradiction. Thus $F(S)$ is an odd integer and Proposition 4.4 ensures that S is symmetric. □

From the above proposition (and its proof) it can be easily seen that symmetric numerical semigroups are those numerical semigroups with type one (see also Exercise 2.5).

Corollary 4.11. *Let S be a numerical semigroup. The following are equivalent.*

1) S is symmetric.
2) $PF(S) = \{F(S)\}$.
3) $t(S) = 1$.

Proof. Observe that $F(S)$ always belongs to $PF(S)$. Thus Conditions 2) and 3) are equivalent. The equivalence between Conditions 1) and 2) follows from the proof of Proposition 4.10. □

Thus in view of Proposition 2.20, Corollary 4.11 can also be reformulated as follows.

Corollary 4.12. *Let S be a numerical semigroup and let n be a nonzero element of S. Then S is symmetric if and only if*

$$\text{Maximals}_{\leq_S} \text{Ap}(S, n) = \{F(S) + n\}.$$

Example 4.13. Let $S = \langle 4, 6, 7 \rangle$. Then $\text{Ap}(S, 4) = \{0, 6, 7, 13\}$. Hence

$$\text{Maximals}_{\leq_S} \text{Ap}(S, 4) = \{13\}$$

and thus $PF(S) = \{9\}$. This means that S is symmetric.

Similar characterizations can be obtained for pseudo-symmetric numerical semigroups, but paying special attention to $\frac{F(S)}{2}$.

Lemma 4.14. *Let S be a pseudo-symmetric numerical semigroup and let n be a positive integer of S. Then $F(S)/2 + n \in \text{Ap}(S, n)$.*

Proof. Since $F(S)/2 \notin S$, we only have to prove that $F(S)/2 + n \in S$. If this were not the case, then by Proposition 4.4, $F(S) - (F(S)/2 + n) = F(S)/2 - n \in S$. But this leads to $F(S)/2 = F(S)/2 - n + n \in S$, which is impossible. □

Observe that this statement is also showing that if S is pseudo-symmetric, then $F(S)/2 \in PF(S)$.

Proposition 4.15. *Let S be a numerical semigroup with even Frobenius number and let $n \in S \setminus \{0\}$. Then S is pseudo-symmetric if and only if*

$$\mathrm{Ap}(S,n) = \{a_0 < a_1 < \cdots < a_{n-2} = F(S) + n\} \cup \left\{\frac{F(S)}{2} + n\right\}$$

and $a_i + a_{n-2-i} = a_{n-2}$ for all $i \in \{0, \ldots, n-2\}$.

Proof. Necessity. By Lemma 4.14, $(F(S)/2) + n \in \mathrm{Ap}(S,n)$. Clearly $(F(S)/2) + n < \max \mathrm{Ap}(S,n) = F(S) + n$ (Proposition 2.12). If $w \in \mathrm{Ap}(S,n) \setminus \{(F(S)/2) + n\}$, then $w - n \notin S$ and $w - n \neq F(S)/2$. By Proposition 4.4, we have that $F(S) - (w - n) \in S$ and thus $\max \mathrm{Ap}(S,n) - w = F(S) + n - w \in S$. By Lemma 4.9, we deduce that $\max \mathrm{Ap}(S,n) - w \in \mathrm{Ap}(S,n)$. Furthermore $\max \mathrm{Ap}(S,n) - w \neq (F(S)/2) + n$ because otherwise we would have $w = F(S)/2$. The proof now follows as the proof of Proposition 4.10.

Sufficiency. Let x be an integer such that $x \neq F(S)/2$ and $x \notin S$. Let us show that $F(S) - x \in S$. Take $w \in \mathrm{Ap}(S,n)$ such that $w \equiv x \bmod n$. Then $x = w - kn$ for some $k \in \mathbb{N} \setminus \{0\}$. We distinguish two cases.

1) If $w = (F(S)/2) + n$, then $F(S) - x = F(S) - ((F(S)/2) + n - kn) = (F(S)/2) + (k-1)n$. Besides, $x \neq F(S)/2$ leads to $k \neq 1$ and therefore $k \geq 2$. Hence we can assert that $F(S) - x \in S$.
2) If $w \neq (F(S)/2) + n$, then $F(S) - x = F(S) - (w - kn) = F(S) + n - w + (k-1)n = a_{n-2} - w + (k-1)n \in S$, since $a_{n-2} - w \in S$ by hypothesis. □

The analogue to Corollary 4.11 for pseudo-symmetric numerical semigroups is stated as follows. As we see with an example we cannot get a condition similar to the third condition in that result.

Corollary 4.16. *Let S be a numerical semigroup. The following conditions are equivalent.*

1) S is pseudo-symmetric.
2) $PF(S) = \{F(S), F(S)/2\}$.

Observe that in this case if $t(S) = 2$, we cannot ensure that $PF(S) = \{\Gamma(S), F(S)/2\}$.

Example 4.17. Let $S = \langle 5, 7, 8 \rangle$. The set of pseudo-Frobenius numbers of S is $PF(S) = \{9, 11\}$. This semigroup has type two, but it is not pseudo-symmetric.

Example 4.18. Let $S = \langle 5, 6, 7, 9 \rangle$. Then $\mathrm{Ap}(S,5) = \{0, 6, 7, 9, 13\}$ and

$$\mathrm{Maximals}_{\leq_S} \mathrm{Ap}(S,5) = \{9, 13\}.$$

This implies that $PF(S) = \{4, 8\} = \{F(S)/2, F(S)\}$. Hence S is pseudo-symmetric.

By using Proposition 2.20, we can obtain an alternative characterization in terms of the Apéry sets.

Corollary 4.19. *Let S be a numerical semigroup and let n be a nonzero element of S. Then S is pseudo-symmetric if and only if*

$$\text{Maximals}_{\leq_S}(\text{Ap}(S,n)) = \left\{ \frac{F(S)}{2} + n, F(S) + n \right\}.$$

2 Irreducible numerical semigroups with arbitrary multiplicity and embedding dimension

We will see that if S is an irreducible numerical semigroup with $m(S) \geq 4$, then $e(S) \leq m(S) - 1$. The aim of this section is to show how to construct, for given m and e integers such that $2 \leq e \leq m-1$, a symmetric numerical semigroup with multiplicity m and embedding dimension e. We already know that numerical semigroups of embedding dimension two are symmetric, thus we cannot find pseudo-symmetric numerical semigroups with embedding dimension two. If we change the above constraint to $3 \leq e \leq m-1$, then we are able to construct a pseudo-symmetric numerical semigroup S with $m(S) = m$ and $e(S) = e$.

2.1 Symmetric case

Lemma 4.20. *Let S be a symmetric numerical semigroup with $m(S) \geq 3$. Then*

$$e(S) \leq m(S) - 1.$$

Proof. Write $\text{Ap}(S,n) = \{0 = a_0 < a_1 < \cdots < a_{m(S)-1}\}$. Then by Proposition 4.10, $a_{m(S)-1} = a_i + a_{m(S)-1-i}$ for all $i \in \{0, \ldots, m(S) - 1\}$. If $m(S) \geq 3$, then we can choose $i = 1$, which implies that $a_{m(S)-1}$ is not a minimal generator. As at least one nonzero element of $\text{Ap}(S,n)$ is not a minimal generator, $e(S) \leq m(S) - 1$ (see Proposition 2.10). □

Remark 4.21. As a consequence of this result and Corollary 4.7, we have that a symmetric numerical semigroup has maximal embedding dimension if and only if it has multiplicity two.

Next we describe a method given in [62] to obtain for fixed integers e and m, with $2 \leq e \leq m-1$, a symmetric numerical semigroup S with $e(S) = e$ and $m(S) = m$.

We introduce two families of symmetric numerical semigroups. Each of them will be used to produce the desired symmetric numerical semigroup depending on the parity of the multiplicity minus the embedding dimension.

Lemma 4.22. *Let m and q be positive integers such that $m \geq 2q+3$ and let S be the submonoid of $(\mathbb{N},+)$ generated by*

$$\{m, m+1, qm+2q+2, \ldots, qm+(m-1)\}.$$

Then S is a symmetric numerical semigroup with multiplicity m, embedding dimension $m-2q$ and Frobenius number $2qm+2q+1$.

Proof. Since $\gcd\{m, m+1\} = 1$, we have that S is a numerical semigroup (by Lemma 2.1). Clearly, $\{m, m+1, qm+2q+2, \ldots, qm+(m-1)\}$ is a minimal system of generators for S and thus $\mathrm{m}(S) = m$ and $\mathrm{e}(S) = m - 2q$. It is easy to deduce that

$$\mathrm{Ap}\,(S,m) = \{0 < m+1 < 2m+2 < \cdots < qm+q < qm+2q+2 < \cdots$$
$$< qm+(m-1) < (q+1)m+q+1 < \cdots < (2q+1)m+2q+1\}.$$

We use Proposition 4.10 to prove that S is symmetric. We must find for all $w \in \mathrm{Ap}(S,m)$ an element $w' \in \mathrm{Ap}(S,m)$ such that $w + w' = (2q+1)m+2q+1$.

- For $i \in \{0, 1, \ldots, m-1-2q-2\}$,

$$(qm+2q+2+i) + (qm+m-1-i) = (2q+1)m+2q+1.$$

- For $k \in \{0, 1, 2, \ldots, q\}$,

$$(km+k) + ((2q+1-k)m+2q+1-k) = (2q+1)m+2q+1.$$

Besides, as $\mathrm{F}(S) + m = \max \mathrm{Ap}\,(S,m)$ (Proposition 2.12), we deduce that $\mathrm{F}(S) = 2qm+2q+1$. □

Lemma 4.23. *Let m and q be nonnegative integers such that $m \geq 2q+4$ and let S be the submonoid of $(\mathbb{N},+)$ generated by*

$$\{m, m+1, (q+1)m+q+2, \ldots, (q+1)m+m-q-2\}.$$

Then S is a symmetric numerical semigroup with multiplicity m, embedding dimension $m-2q-1$ and Frobenius number $2(q+1)m-1$.

Proof. As in Lemma 4.22, we deduce that S is a numerical semigroup with $\mathrm{m}(S) = m$ and $\mathrm{e}(S) = m - 2q - 1$.

It is easy to prove that

$$\mathrm{Ap}\,(S,m) = \{0 < m+1 < 2m+2 < \cdots$$
$$< (q+1)m+q+1 < (q+1)m+q+2 < \cdots < (q+1)m+m-q-2$$
$$< (q+2)m+m-q-1 < (q+3)m+(m-q) < \cdots < 2(q+1)m+m-1\}.$$

Furthermore, as a consequence of the following comments, by Proposition 4.10, S is symmetric.

- For $i \in \{0, 1, \ldots, m - 2q - 3\}$,

$$((q+1)m+q+1+i) + ((q+1)m+m-q-2-i) = 2(q+1)m+m-1.$$

- For $k \in \{0, 1, \ldots, q\}$,

$$((q+2+k)m+m-q-1-k) + ((q-k)m+q-k) = 2(q+1)m+m-1.$$

As $F(S) + m = \max(\mathrm{Ap}(S, m))$ (see Proposition 2.12), we obtain that $F(S) = 2(q+1)m - 1$. $\square$

Theorem 4.24. *Let m and e be integers such that $2 \leq e \leq m - 1$. There exists a symmetric numerical semigroup with multiplicity m and embedding dimension e.*

Proof. If $e = 2$, then $S = \langle m, m+1 \rangle$ is a symmetric numerical semigroup with multiplicity m and embedding dimension 2 (Corollary 4.7). Thus, in the sequel, we may assume that $e \geq 3$. We distinguish two cases.

- If $m - e$ is even, then there exists $q \in \mathbb{N} \setminus \{0\}$ such that $m - e = 2q$. Furthermore, $e \geq 3$ implies that $m \geq m - e + 3$ and therefore $m \geq 2q + 3$. Lemma 4.22 ensures the existence of a symmetric numerical semigroup with multiplicity m and embedding dimension $e = m - 2q$.
- If $m - e$ is odd, then there exists $q \in \mathbb{N}$ such that $m - e = 2q + 1$. The constraint $e \geq 3$ implies that $m \geq m - e + 3$ and thus $m \geq 2q + 4$. Lemma 4.23 is used now to construct a symmetric numerical semigroup with multiplicity m and embedding dimension $e = m - 2q - 1$. $\square$

Example 4.25 ([62]). The semigroup $\langle 12, 13, 44, 45, 46, 47 \rangle$ is symmetric with multiplicity 12, embedding dimension 6 and Frobenius number 79 ($q = 3$).

The semigroup $\langle 15, 16, 81, 82, 83, 84 \rangle$ is symmetric and has multiplicity 15, embedding dimension 6 and Frobenius number 149 ($q = 4$).

2.2 Pseudo-symmetric case

We now proceed with the pseudo-symmetric case. In view of Lemma 4.14 we will encounter slight differences with the symmetric case. The construction we explain in this section appears in [74].

We start by proving that for multiplicity greater than or equal to four, the embedding dimension of a pseudo-symmetric numerical semigroup never reaches the multiplicity.

Lemma 4.26 ([73]). *Let S be a pseudo-symmetric numerical semigroup with $\mathrm{m}(S) \geq 4$. Then*

$$e(S) \leq \mathrm{m}(S) - 1.$$

Proof. By Proposition 2.10, $e(s) \leq m(S)$. If $e(S) = m(S)$, then S is minimally generated by $\{m(S), n_1, \ldots, n_{m(S)-1}\}$ and by Proposition 4.15, $\text{Ap}(S, m(S))$ is of the form

$$\text{Ap}(S, m(S)) = \{0 < n_2 < \cdots < n_{m(S)-1}\} \cup \left\{ n_1 = \frac{F(S)}{2} + m(S) \right\}.$$

As $m(S) - 1 \geq 3$, by Proposition 4.15, we deduce that $n_{m(S)-1} - n_2 \in S$, which contradicts the fact that $\{m(S), n_1, \ldots, n_{m(S)-1}\}$ is a minimal system of generators for S. $\qquad\square$

In view of this result, we must pay special attention to the case of multiplicity less than four. Numerical semigroups with multiplicity two are symmetric (as a consequence of Corollary 4.7). Thus we must study those with multiplicity three.

Lemma 4.27 ([73]). *The following conditions are equivalent.*

1) S is a pseudo-symmetric numerical semigroup with $m(S) = e(S) = 3$.
2) $S = \langle 3, x+3, 2x+3 \rangle$ with x an integer not divisible by three.

Proof. 1) implies 2). If $m(S) = e(S) = 3$, then $\{3, n_1, n_2\}$ is a minimal system of generators for S. From Proposition 4.4, we deduce that $F(S)$ is even, and by Proposition 4.15 we have that

$$\text{Ap}(S, 3) = \left\{ 0, n_1 = \frac{F(S)}{2} + 3, n_2 = F(S) + 3 \right\}.$$

By taking $x = F(S)/2$ we have that $n_1 = x+3$ and $n_2 = 2x+3$. Since $x = F(S)/2 \notin S$, we get that x is not a multiple of 3.

2) implies 1). Clearly $\{3, x+3, 2x+3\}$ is a minimal system of generators for S, and thus $m(S) = e(S) = 3$. Hence $\text{Ap}(S, 3) = \{0, x+3, 2x+3\}$. By Proposition 2.12, $2x+3 = F(S) + 3$, and consequently $(F(S)/2) + 3 = x+3$. Proposition 4.15 asserts that S is pseudo-symmetric. $\qquad\square$

By Corollary 4.7, every embedding dimension two numerical semigroup is symmetric. We see that for embedding dimension three, there are always pseudo-symmetric numerical semigroups with arbitrary multiplicity.

Lemma 4.28. *Let m be a positive integer greater than or equal to four. There exists a pseudo-symmetric numerical semigroup S with $F(S)$ even, $m(S) = m$ and $e(S) = 3$.*

Proof. We distinguish two cases depending on the parity of m.

1) If m is even, then $m = 2q + 4$ for some $q \in \mathbb{N}$. Let

$$S = \langle m, m+1, (q+1)m + (m-1) \rangle.$$

It is clear that $m(S) = m$ and $e(S) = 3$. Under this condition

$$\text{Ap}(S, m) = \{0, m+1, 2(m+1), \ldots, (m-2)(m+1)\} \cup \{(q+1)m + (m-1)\}.$$

By Proposition 2.12, $F(S) = (m-2)m-2$, which is even, and $\frac{F(S)}{2} + m = (q+1)m + (m-1)$. By Proposition 4.15 we conclude that S is an irreducible numerical semigroup.

2) If m is odd, then $m = 2q+3$ for some $q \in \mathbb{N} \setminus \{0\}$. Let

$$S = \langle m, m+1, (q+1)m+q+2 \rangle.$$

Clearly, $\mathrm{m}(S) = m$ and $\mathrm{e}(S) = 3$. In this setting,

$$\mathrm{Ap}(S,m) = \{0, m+1, 2(m+1), \ldots, q(m+1), (q+1)m+q+2,$$
$$(m+1) + (q+1)m+q+2, \ldots, q(m+1) + (q+1)m+q+2\}$$
$$\cup \{(q+1)(m+1)\}.$$

Hence, $F(S) = 2(1+q+mq)$ is even and $\frac{F(S)}{2} + m = (q+1)(m+1)$. By Proposition 4.15, we have that S is pseudo-symmetric. $\qquad\square$

We now proceed as in the symmetric case by presenting two families of pseudo-numerical semigroups that will be used depending on the parity of the desired multiplicity minus the desired embedding dimension.

Lemma 4.29. *Let $m,q \in \mathbb{N}$ be such that $m \geq 2q+5$ and let S be the submonoid of $(\mathbb{N}, +)$ generated by*

$$\{m, m+1, (q+1)m+q+2, \ldots, (q+1)m+m-q-3, (q+1)m+m-1\}.$$

Then S is a pseudo-symmetric numerical semigroup with $\mathrm{m}(S) = m$, $\mathrm{e}(S) = m - 2q-1$ and $F(S) = 2(q+1)m-2$.

Proof. Since $\gcd\{m, m+1\} = 1$, we have that S is a numerical semigroup (by Lemma 2.1). Note that $m = \min S \setminus \{0\}$ and so $\mathrm{m}(S) = m$. It is straightforward to see that

$$\{n_0 = m, n_1 = m+1, n_2 = (q+1)m+q+2, \ldots,$$
$$n_{p-1} = (q+1)m+m-q-3, n_p = (q+1)m+m-1\}$$

is a minimal system of generators for S and thus $\mathrm{e}(S) = m-2q-1$. It is easy to check that

$$\mathrm{Ap}(S,m) = \{0, n_1, 2n_1, \ldots, (q+1)n_1, n_2, \ldots, n_{p-1}, n_1 + n_{p-1}, 2n_1 + n_{p-1}, \ldots,$$
$$qn_1 + n_{p-1}, F(S) + m = (q+1)n_1 + n_{p-1}\} \cup \{n_p\},$$

and if $p \geq 4$, then in addition $F(S) + m = n_i + n_{p-i}$ for all $i \in \{2, \ldots, \lceil p/2 \rceil\}$. Hence, $F(S) = 2(q+1)m-2$ and so $\frac{F(S)}{2} + m = (q+1)m + (m-1) = n_p$. By applying Proposition 4.15, we deduce that S is pseudo-symmetric. $\qquad\square$

Lemma 4.30. *Let $m \in \mathbb{N}$ and $q \in \mathbb{N} \setminus \{0\}$ be such that $m \geq 2q+4$ and let S be the submonoid of $(\mathbb{N}, +)$ generated by*

$$\{m, m+1, qm+2q+3, \ldots, qm+m-1, (q+1)m+q+2\}.$$

Then S is a pseudo-symmetric numerical semigroup with $\mathrm{m}(S) = m$, $\mathrm{e}(S) = m-2q$ and $\mathrm{F}(S) = 2qm+2q+2$.

Proof. As $\gcd\{m, m+1\} = 1$, S is a numerical semigroup (as a consequence of Lemma 2.1). Since $m = \min S \setminus \{0\}$, we get that $\mathrm{m}(S) = m$. Clearly,

$$\{n_0 = m, n_1 = m+1, n_2 = qm+2q+3, \ldots,$$
$$n_{p-1} = qm+(m-1), n_p = (q+1)m+q+2\}$$

is a minimal system of generators for S and so $\mathrm{e}(S) = m - 2q$. The reader can prove that

$$\mathrm{Ap}(S, m) = \{0, n_1, 2n_1, \ldots, qn_1, n_2, \ldots, n_{p-1}, n_p, n_1 + n_p, 2n_1 + n_p, \ldots,$$
$$\mathrm{F}(S) + m = qn_1 + n_p\} \cup \{(q+1)n_1\},$$

and $\mathrm{F}(S) + m = n_i + n_{p-i+1}$ for all $i \in \{2, \ldots, \lceil (p+1)/2 \rceil\}$. Then $\mathrm{F}(S) = 2qm + 2q + 2$ and thus $\frac{\mathrm{F}(S)}{2} + m = (q+1)m + q + 1 = (q+1)n_1$. Proposition 4.15 asserts that S is a pseudo-symmetric numerical semigroup.   □

Theorem 4.31. *Let m and e be positive integers such that $3 \leq e \leq m - 1$. Then there exists a pseudo-symmetric numerical semigroup with multiplicity m and embedding dimension e.*

Proof. If $e = 3$, then Lemma 4.28 ensures the existence of this semigroup. Thus, in sequel, we shall assume that $4 \leq e \leq m - 1$. We distinguish two cases depending on the parity of $m - e$.

1) If $m - e$ is odd, then there exists $q \in \mathbb{N}$ such that $m - e = 2q + 1$. Moreover, since $e \geq 4$, $m \geq 2q + 5$. By Lemma 4.29, we deduce that there exists a pseudo-symmetric numerical semigroup S with $\mathrm{m}(S) = m$ and $\mathrm{e}(S) = m - 2q - 1 = e$.
2) If $m - e$ is even, then there exists $q \in \mathbb{N} \setminus \{0\}$ such that $m - e = 2q$. As $e \geq 4$, $m \geq 2q + 4$. By Lemma 4.30, we deduce that there exists a pseudo-symmetric numerical semigroup S with $\mathrm{m}(S) = m$ and $\mathrm{e}(S) = m - 2q = e$.   □

Example 4.32 ([74]). The numerical semigroup $S = \langle 11, 12, 37, 38, 39, 43 \rangle$ is pseudo-symmetric with $\mathrm{m}(S) = 11$, $\mathrm{e}(S) = 6$ and $\mathrm{F}(S) = 64$ ($q = 2$).

The semigroup $S = \langle 11, 12, 29, 30, 31, 32, 37 \rangle$ is a pseudo-symmetric numerical semigroup with $\mathrm{m}(S) = 11$, $\mathrm{e}(S) = 7$ and $\mathrm{F}(S) = 50$ ($q = 2$).

As for embedding dimension three,

- $S = \langle 6, 7, 17 \rangle$ is an irreducible numerical semigroup with $\mathrm{m}(S) = 6$ and $\mathrm{F}(S) = 22$.
- $S = \langle 7, 8, 25 \rangle$ is an irreducible numerical semigroup with $\mathrm{m}(S) = 7$ and $\mathrm{F}(S) = 34$.

3 Unitary extensions of a numerical semigroup

We introduce the concept of special gap of a numerical semigroup. Its definition is motivated by the problem of finding the set of all numerical semigroups containing a given numerical semigroup.

Given a numerical semigroup S, denote by

$$SG(S) = \{x \in PF(S) \mid 2x \in S\}.$$

Its elements will be called the *special gaps* of S.

It is easy to prove that the elements of $SG(S)$ are precisely those gaps x of S such that $S \cup \{x\}$ is again a numerical semigroup.

Proposition 4.33. *Let S be a numerical semigroup and let $x \in G(S)$. The following properties are equivalent:*

(1) $x \in SG(S)$,
(2) $S \cup \{x\}$ is a numerical semigroup.

Example 4.34. Let $S = \{0, 7, \rightarrow\}$. Then S is a numerical semigroup with $PF(S) = \{1,2,3,4,5,6\}$, and consequently $SG(S) = \{4,5,6\}$. This implies that $\{0,4,7,\rightarrow\}$, $\{0,5,7,\rightarrow\}$ and $\{0,6,\rightarrow\}$ are numerical semigroups.

If the numerical semigroup S is properly contained in a numerical semigroup T and we take $x = \max(T \setminus S)$, then $x + s \in T$ and $x + s > x$ for all $s \in S^*$. Thus $x + s \in S$. Analogously, $2x \in T$ and $2x > x$, which implies that $2x \in S$. This proves the following result.

Lemma 4.35. *Let S and T be two numerical semigroups such that $S \subsetneq T$. Then $S \cup \{\max(T \setminus S)\}$ is a numerical semigroup, or equivalently, $\max(T \setminus S) \in SG(S)$.*

Given a numerical semigroup S, we denote by $\mathscr{O}(S)$ the set of all numerical semigroups that contain S. We will refer to $\mathscr{O}(S)$ as the set of *oversemigroups* of S. Since the complement of S in $\mathbb{N}$ is finite, $\mathscr{O}(S)$ is finite.

Given two numerical semigroups S and T with $S \subseteq T$, we define recursively

- $S_0 = S$,
- $S_{n+1} = S_n \cup \{\max(T \setminus S_n)\}$ if $S_n \neq T$, and $S_n = S_{n+1}$ otherwise.

If the cardinality of $T \setminus S$ is k, then

$$S = S_0 \subsetneq S_1 \subsetneq \cdots \subsetneq S_k = T.$$

By using this idea we can construct the set $\mathscr{O}(S)$. We start setting $\mathscr{O}(S) = \{S\}$, and then for every element in $\mathscr{O}(S)$ not equal to $\mathbb{N}$ (observe that $SG(\mathbb{N})$ is the empty set), we attach to $\mathscr{O}(S)$ the numerical semigroups $S \cup \{x\}$ with x ranging in $SG(S)$.

Example 4.36 ([90]). Let $S = \langle 5,7,9,11 \rangle$. For this semigroup, $\mathrm{SG}(S) = \{13\}$ and thus $S \cup \{13\} = \langle 5,7,9,11,13 \rangle$ is a semigroup containing S (the only one that differs in just one element). As $\mathrm{SG}(S \cup \{13\}) = \{6,8\}$, from $S \cup \{13\}$ we obtain two new semigroups which are $S \cup \{13,6\}$ and $S \cup \{13,8\}$. By repeating this process we obtain $\mathscr{O}(S)$, which we draw below as a graph.

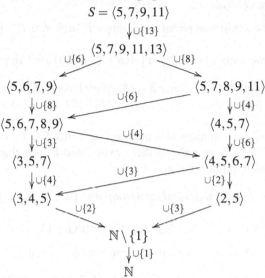

As a consequence of Lemma 4.35, if S is a numerical semigroup, then S is maximal (with respect to set inclusion) in the set of all numerical semigroups not cutting $\mathrm{SG}(S)$. Moreover, $\mathrm{SG}(S)$ is the smallest set of gaps that determines S up to maximality.

Proposition 4.37. *Let S be a numerical semigroup and let $\{g_1,\dots,g_t\} \subseteq \mathrm{G}(S)$. The following conditions are equivalent.*

1) S is maximal (with respect to set inclusion) in the set of all numerical semigroups T such that $T \cap \{g_1,\dots,g_t\}$ is empty.

2) $\mathrm{SG}(S) \subseteq \{g_1,\dots,g_t\}$.

Proof. Let $x \in \mathrm{SG}(S)$. By Proposition 4.33, $S \cup \{x\}$ is a numerical semigroup containing S properly. Thus if Condition 1) holds, then $(S \cup \{x\}) \cap \{g_1,\dots,g_t\} \neq \emptyset$. Hence $x \in \{g_1,\dots,g_t\}$.

The implication *2) implies 1)* follows easily from Proposition 4.33 and Lemma 4.35. $\qquad\square$

As a corollary we find another characterization of irreducible numerical semigroups. By Theorem 4.2, we know that a numerical semigroup S is irreducible if and only if it is maximal in the set of numerical semigroups that do not cut $\{F(S)\}$. Clearly $F(S)$ belongs to $\mathrm{SG}(S)$, whenever S is not equal to $\mathbb{N}$.

Corollary 4.38. *Let S be a numerical semigroup. Then S is irreducible if and only if $\mathrm{SG}(S)$ has at most one element.*

The following two results enable us to find a generalization of the construction proposed in Lemma 4.3 to find an irreducible oversemigroup of a given numerical semigroup with the same Frobenius number.

Lemma 4.39. *Let S be a numerical semigroup and let $\{g_1,\ldots,g_t\} \subseteq G(S)$. The following conditions are equivalent.*

1) *S is maximal in the set of numerical semigroups T such that $T \cap \{g_1,\ldots,g_t\}$ is empty.*
2) *If $x \in G(S)$, then there exist $i \in \{1,\ldots,t\}$ and $k \in \mathbb{N} \setminus \{0\}$ such that $g_i - kx \in S$.*

Proof. 1) *implies* 2). Let $x \in G(S)$. Since $S \subsetneq \langle S, x \rangle$, we have that $\langle S, x \rangle \cap \{g_1,\ldots,g_t\}$ is not empty. Hence there exist $i \in \{1,\ldots,t\}$, $k \in \mathbb{N} \setminus \{0\}$ and $s \in S$ such that $g_i = s + kx$. This leads to $g_i - kx \in S$.

2) *implies* 1). Let T be a numerical semigroup such that $S \subsetneq T$. Take $x \in T \setminus S$. Then $S \subsetneq \langle S, x \rangle \subseteq T$ and by hypothesis there exist i and k such that $g_i - kx \in S$. Hence $g_i \in \langle S, x \rangle$, which implies that $g_i \in T$. $\qquad\square$

Proposition 4.40. *Let S be a numerical semigroup and $\{g_1,\ldots,g_t\} \subseteq G(S)$. If there exists*

$$h = \max \{x \in \mathbb{Z} \setminus S \mid 2x \in S, g_i - x \notin S \text{ for all } i \in \{1,\ldots,t\}\},$$

then $S \cup \{h\}$ is a numerical semigroup not intersecting $\{g_1,\ldots,g_t\}$.

Proof. By Proposition 4.33, we must prove that $h \in SG(S)$. Clearly $2h \in S$. Assume that there exists $s \in S \setminus \{0\}$ such that $h + s \notin S$. Since $2(h+s) \in S$ and $h < h+s$, we get that $g_i - (h+s) \in S$ for some $i \in \{1,\ldots,t\}$. But this yields $g_i - h \in S$, in contradiction with the definition of h. $\qquad\square$

Corollary 4.41. *Let S be a numerical semigroup and $\{g_1,\ldots,g_t\} \subseteq G(S)$. The following conditions are equivalent.*

1) *S is maximal in the set of all numerical semigroups whose intersection with $\{g_1,\ldots,g_t\}$ is empty.*
2) *For every $x \in \mathbb{N}$, if $x \in G(S)$ and $2x \in S$, then $g_i - x \in S$ for some $i \in \{1,\ldots,t\}$.*

Proof. 1) *implies* 2) follows from Proposition 4.40.

For the other implication, in view of Lemma 4.39, it suffices to show that for every $x \in G(S)$, there exist appropriate i and k such that $g_i - kx \in S$. Let $x \in G(S)$ and set $k = \max \{n \in \mathbb{N} \setminus \{0\} \mid nx \notin S\}$. Clearly $kx \in G(S)$ and $2kx \in S$. By hypothesis $g_i - kx \in S$ for some $i \in \{1,\ldots,t\}$. $\qquad\square$

As a consequence of these two results, we obtain a characterization of irreducible numerical semigroups that gathers Conditions 1) and 2) of Proposition 4.4.

Corollary 4.42. *Let S be a numerical semigroup. Then S is irreducible if and only if for all $x \in \mathbb{N}$, $x \in G(S)$ and $2x \in S$ imply $F(S) - x \in S$.*

Proof. Follows from Theorem 4.2 and Corollary 4.41. $\qquad\square$

Let $\mathscr{S}$ be the set of all numerical semigroups. For $\{g_1,\dots,g_t\}\subset\mathbb{N}$, define

$$\mathscr{S}(g_1,\dots,g_t)=\big\{S'\in\mathscr{S}\mid S'\cap\{g_1,\dots,g_t\}=\emptyset\big\}.$$

Proposition 4.40 can be used to find a maximal element in $\mathscr{S}(g_1,\dots,g_t)$. We only have to take as starting point $S=\{0,\max\{g_1,\dots,g_t\}+1,\to\}$ and define recursively

- $S_0=S$,
- $S_{n+1}=S_n\cup\{\mathrm{h}(S_n)\}$, where $\mathrm{h}(S_n)$ is

$$\max\big\{x\in G(S_n)\mid 2x\in S_n, g_i-x\notin S_n \text{ for all } i\in\{1,\dots,t\}\big\};$$

if $\mathrm{h}(S_n)$ does not exist, then S_n is the desired semigroup (Corollary 4.41 gives this stop condition).

Example 4.43 ([90]). We compute an element in $\mathrm{Maximals}_\subseteq(\mathscr{S}(5,6))$.

(1) $S_0=S=\{0,7,\to\}$, $\mathrm{h}(S_0)=4$,
(2) $S_1=\{0,4,7,\to\}$, $\mathrm{h}(S_1)$ does not exist and thus S_1 belongs to $\mathrm{Maximals}_\subseteq(\mathscr{S}(5,6))$.

4 Decomposition of a numerical semigroup into irreducibles

We present a procedure to compute a decomposition of a given numerical semigroup into irreducible numerical semigroups. We will also show how to obtain "minimal" decompositions.

Let S be a numerical semigroup. If S is not irreducible, then there exists S_1 and S_2 properly containing it such that $S=S_1\cap S_2$. We might wonder now if S_1 or S_2 are irreducible, and in the negative write them as an intersection of two other numerical semigroups. We can repeat several times this process, but only a finite number of times, since every numerical semigroup appearing in this procedure properly contains S and $\mathscr{O}(S)$ is finite.

Proposition 4.44. *Every numerical semigroup can be expressed as the intersection of finitely many irreducible numerical semigroups.*

Recall that we have a procedure to construct $\mathscr{O}(S)$ for any numerical semigroup S, based on the computation of the set $SG(S)$. While performing this procedure we can choose those over semigroups with at most one special gap, which in view of Corollary 4.38 are those irreducible oversemigroups of S. Denote by

$$\mathscr{I}(S)=\{T\in\mathscr{O}(S)\mid T \text{ is irreducible}\}.$$

It follows that $S=\bigcap_{T\in\mathscr{I}(S)}T$. We can remove from this intersection those elements that are not minimal with respect to set inclusion, and the resulting semigroup remains unchanged.

Proposition 4.45. *Let S be a numerical semigroup and let*

$$\{S_1,\dots,S_n\} = \text{Minimals}_\subseteq \mathscr{I}(S).$$

Then

$$S = S_1 \cap \cdots \cap S_n.$$

This decomposition does not have to be minimal (in the sense of minimal number of irreducibles involved) as the following example shows.

Example 4.46 ([90]). Let $S = \langle 5,6,8 \rangle$. We compute the set $\text{Minimals}_\subseteq \mathscr{I}(S)$. Since $\text{EH}(S) = \{7,9\}$, by Proposition 4.33, $S \cup \{7\}$ and $S \cup \{9\}$ are numerical semigroups. As $\text{SG}(S \cup \{7\}) = \{9\}$, $S \cup \{7\}$ is irreducible (Corollary 4.38), which implies that it belongs to $\text{Minimals}_\subseteq(\mathscr{I}(S))$. The semigroup $S \cup \{9\}$ is not irreducible ($\text{SG}(S \cup \{9\}) = \{3,4,7\}$). By Proposition 4.33 the sets $S \cup \{9,3\}$, $S \cup \{9,7\}$ and $S \cup \{9,4\}$ are also numerical semigroups. Both $S \cup \{9,3\}$ and $S \cup \{9,4\}$ are irreducible semigroups, and $S \cup \{9,7\}$ contains the semigroup $S \cup \{7\}$ (the first irreducible we have found). Hence the set

$$\text{Minimals}_\subseteq(\mathscr{I}(S)) = \{S \cup \{7\}, S \cup \{9,3\}, S \cup \{9,4\}\}.$$

Finally,

$$S = (S \cup \{7\}) \cap (S \cup \{9,4\}) \cap (S \cup \{9,3\}) = (S \cup \{7\}) \cap (S \cup \{9,4\}). \qquad \square$$

When looking for the least n such that $S = S_1 \cap \cdots \cap S_n$, with $S_1,\dots,S_n \in \mathscr{I}(S)$, then it suffices to search among the decompositions with elements in $\text{Minimals}_\subseteq(\mathscr{I}(S))$.

Proposition 4.47. *Let S be a numerical semigroup. If $S = S_1 \cap \cdots \cap S_n$ with $S_1,\dots,S_n \in \mathscr{I}(S)$, then there exists $S_1',\dots,S_n' \in \text{Minimals}_\subseteq(\mathscr{I}(S))$ such that*

$$S = S_1' \cap \cdots \cap S_n'.$$

Proof. For every $i \in \{1,\dots,n\}$, if S_i does not belong to $\text{Minimals}_\subseteq(\mathscr{I}(S))$, then take $S_i' \in \text{Minimals}_\subseteq(\mathscr{I}(S))$ such that $S_i' \subseteq S_i$. $\qquad \square$

The next proposition gives a clue on which semigroups must appear in a minimal decomposition.

Proposition 4.48. *Let S be a numerical semigroup and let $S_1,\dots,S_n \in \mathscr{O}(S)$. The following conditions are equivalent.*

1) $S = S_1 \cap \cdots \cap S_n$.
2) For all $h \in \text{SG}(S)$, there exists $i \in \{1,\dots,n\}$ such that $h \notin S_i$.

Proof. 1) implies 2). If $h \in \text{SG}(S)$, then $h \notin S$ and thus $h \notin S_i$ for some $i \in \{1,\dots,n\}$.
2) implies 1). If $S \subsetneq S_1 \cap \cdots \cap S_n$, then by Lemma 4.35, $h = \max((S_1 \cap \cdots \cap S_n) \setminus S)$ is in $\text{SG}(S)$, and in all the S_i, in a contradiction with the hypothesis. $\qquad \square$

We can compute $\text{Minimals}_{\subseteq}(\mathscr{I}(S)) = \{S_1, \ldots, S_n\}$. For every $i \in \{1, \ldots, n\}$, set

$$C(S_i) = \{h \in \text{SG}(S) \mid h \notin S_i\}.$$

By Proposition 4.48 we know that

$$S = S_{i_1} \cap \cdots \cap S_{i_r} \text{ if and only if } C(S_{i_1}) \cup \cdots \cup C(S_{i_r}) = \text{SG}(S).$$

From the above results we can obtain a method for computing a decomposition of S as an intersection of irreducible semigroups with the least possible number of them.

For a set Y, we use $\#Y$ to denote its cardinality.

Algorithm 4.49. Let S be a non-irreducible semigroup.

(1) Compute the set $\text{SG}(S)$.
(2) Set $I = \emptyset$ and $C = \{S\}$.
(3) For all $S' \in C$, compute (using Proposition 4.33) all the semigroups $\overline{S}$ such that $\#(\overline{S} \setminus S') = 1$. Remove S' from C. Let B be the set formed by the semigroups constructed in this way.
(4) Remove from B the semigroups S' fulfilling that $\text{SG}(S) \subseteq S'$.
(5) Remove from B the semigroups S' such that there exists $\tilde{S} \in I$ with $\tilde{S} \subseteq S'$.
(6) Set $C = \{S' \in B \mid S' \text{ is not irreducible}\}$.
(7) Set $I = I \cup \{S' \in B \mid S' \text{ is irreducible}\}$.
(8) If $C \neq \emptyset$, go to Step 3.
(9) For every $S \in I$, compute $C(S)$.
(10) Choose $\{S_1, \ldots, S_r\}$ such that r is minimum fulfilling that

$$C(S_1) \cup \cdots \cup C(S_r) = \text{SG}(S).$$

(11) Return $S_1, \ldots, S_r$.

Next we illustrate this method with an example.

Example 4.50. We take again the semigroup $S = \langle 5, 6, 8 \rangle$. We have that $\text{SG}(S) = \{7, 9\}$. Performing the steps of the above algorithm we get (in Steps 6 and 7) that $I = \{\langle 5, 6, 7, 8 \rangle\}$ and $C = \{\langle 5, 6, 8, 9 \rangle\}\}$. Since $C \neq \emptyset$, we go back to Step 3 obtaining that $I = \{\langle 5, 6, 7, 8 \rangle, \langle 3, 5 \rangle, \langle 4, 5, 6 \rangle\}$ and $C = \emptyset$. Step 8 yields

$$C(\langle 5, 6, 7, 8 \rangle) = \{9\}, \ C(\langle 3, 5 \rangle) = \{7\}, \ C(\langle 4, 5, 6 \rangle) = \{7\}.$$

The minimal decompositions of S are

$$S = \langle 5, 6, 7, 8 \rangle \cap \langle 3, 5 \rangle$$

and

$$S = \langle 5, 6, 7, 8 \rangle \cap \langle 4, 5, 6 \rangle.$$

Given a numerical semigroup we can consider two kinds of minimality in a decomposition of this semigroup into irreducibles. The first is in terms of the cardinality, that is *minimal* in the sense that the least possible number of irreducibles appear in the decomposition. The second is in terms of redundancy, that is, a decomposition is *minimal* if no semigroup involved is redundant (cannot be eliminated and the intersection remains the same), or in other words, it cannot be refined into a smaller decomposition. Both concepts do not coincide. Clearly a decomposition with the least possible number of irreducibles involved cannot be refined. However there are decompositions that cannot be refined with more irreducibles than other decompositions.

Example 4.51. The numerical semigroup $S = \langle 5, 21, 24, 28, 32 \rangle$ can be expressed as

$$S = \langle 5, 9, 12, 13 \rangle \cap \langle 5, 11, 12, 13 \rangle \cap \langle 5, 12, 14, 16 \rangle \cap \langle 5, 14, 16, 18 \rangle$$

and as

$$S = \langle 5, 7 \rangle \cap \langle 5, 8 \rangle.$$

Let us make the computations with the numericalsgps package.

```
gap> s:=NumericalSemigroup(5,21,24,28,32);
<Numerical semigroup with 5 generators>
gap> DecomposeIntoIrreducibles(s);
[ <Numerical semigroup>, <Numerical semigroup>,
<Numerical semigroup>, <Numerical semigroup> ]
gap> l:=last;;
gap> List(l,
> MinimalGeneratingSystemOfNumericalSemigroup);
[ [ 5, 9, 12, 13 ], [ 5, 11, 12, 13 ],
[ 5, 12, 14, 16 ], [ 5, 14, 16, 18 ] ]
gap> s=IntersectionOfNumericalSemigroups(
>NumericalSemigroup(5,7),NumericalSemigroup(5,8));
true
gap> s=IntersectionOfNumericalSemigroups(l[1],
> IntersectionOfNumericalSemigroups(l[2],l[3]));
false
gap> s=IntersectionOfNumericalSemigroups(l[1],
> IntersectionOfNumericalSemigroups(l[2],l[4]));
false
gap> s=IntersectionOfNumericalSemigroups(l[1],
> IntersectionOfNumericalSemigroups(l[3],l[4]));
false
gap> s=IntersectionOfNumericalSemigroups(l[2],
> IntersectionOfNumericalSemigroups(l[3],l[4]));
false
```

5 Fundamental gaps of a numerical semigroup

In view of Proposition 4.37, if S is a numerical semigroup, the set $SG(S)$ determines S up to maximality (with respect to set inclusion). This does not mean that it determines it uniquely. There can be found numerical semigroups S and T with $T \neq S$ and $SG(S) = SG(T)$ (this implies that neither $S \subseteq T$ nor $T \subseteq S$).

Example 4.52. For S in $\{\langle 3,8,13 \rangle, \langle 4,7,9 \rangle, \langle 6,7,8,9,11 \rangle\}$, $SG(S) = \{10\}$.

We present in this section a subset of the set of gaps of a numerical semigroup that fully determines it. This subset was introduced in [91]. Most of the results appearing in this section can be found there.

Let S be a numerical semigroup. We say that a set X of positive integers determines the gaps of S if S is the maximum (with respect to set inclusion) numerical semigroup such that $X \subseteq G(S)$.

Given $X \subseteq \mathbb{N}$ we denote by $D(X)$ the set of all positive divisors of the elements of X, that is,

$$D(X) = \{\, a \in \mathbb{N} \mid a \text{ divides some } x \in X \,\}.$$

Proposition 4.53. *Let X be a finite set of positive integers. The following conditions are equivalent.*

1) The set X determines the gaps of a numerical semigroup.
2) $\mathbb{N} \setminus D(X)$ is a numerical semigroup.

If these conditions hold, then X determines the gaps of the numerical semigroup $\mathbb{N} \setminus D(X)$.

Proof. 1) implies 2). Let S be the numerical semigroup whose gaps are determined by X. Since $X \subseteq G(S)$, we have that $D(X) \subseteq G(S)$ and thus $S \subseteq \mathbb{N} \setminus D(X)$. Take $a \in \mathbb{N} \setminus D(X)$. Then $S' = \langle a, \max(X) + 1, \rightarrow \rangle$ is a numerical semigroup such that $X \subseteq G(S')$, and from the definition of S, we have that $S' \subseteq S$. Hence $a \in S$ and this proves that $\mathbb{N} \setminus D(X) = S$. In particular we obtain that $\mathbb{N} \setminus D(X)$ is a numerical semigroup.

2) implies 1). Obviously, $\mathbb{N} \setminus D(X)$ is the numerical semigroup whose gaps are determined by X. $\qquad \square$

Proposition 4.54. *Let S be a numerical semigroup and let X be a subset of $G(S)$. The following conditions are equivalent.*

1) X determines the gaps of S.
2) For every $a \in \mathbb{N}$, if $a \in G(S)$ and $\{2a, 3a\} \subset S$, then $a \in X$.

Proof. If X determines the gaps of S, then by applying Proposition 4.53, we have that $S = \mathbb{N} \setminus D(X)$, and consequently $G(S) = D(X)$. If a is an element of $G(S)$, then there exists $x \in X$ such that $a \mid x$ and thus $ka = x$ for some $k \in \mathbb{N}$. If in addition we assume that $\{2a, 3a\} \subset S$, we have that $la \in S$ for every positive integer l greater than one. Therefore $k = 1$ and $a = x \in X$.

For the other implication, in view of Proposition 4.53, it suffices to prove that $S = \mathbb{N} \setminus D(X)$. By hypothesis $X \subseteq G(S)$ and thus $D(X) \subseteq G(S)$. Hence $S \subseteq \mathbb{N} \setminus D(X)$. If a is a nonnegative integer not belonging to S, then $a \in G(S)$. Let $k = \max \{ n \in \mathbb{N} \mid na \in G(S) \}$ ($G(S)$ is finite, $0 \notin G(S)$ and thus $k \in \mathbb{N} \setminus \{0\}$). It follows that $ka \in G(S)$ and $\{2ka, 3ka\} \subset S$. This implies by hypothesis that $ka \in X$, and consequently $a \in D(X)$. This proves $S = \mathbb{N} \setminus D(X)$.   □

This result motivates the following definition. A gap x of a numerical semigroup S is *fundamental* if $\{2x, 3x\} \subset S$ (or equivalently, $kx \in S$ for all $k > 1$). We denote by $FG(S)$ the set of fundamental gaps of S.

Example 4.55. Let $S = \langle 5, 8, 9 \rangle = \{0, 5, 8, 9, 10, 13, \rightarrow\}$. Then $FG(S) = \{7, 11, 12\}$.

With this new notation we can reformulate Proposition 4.54.

Corollary 4.56. *Let S be a numerical semigroup and let X be a subset of $G(S)$. Then X determines the gaps of S if and only if $FG(S) \subseteq X$.*

Hence for a numerical semigroup S, $FG(S)$ is just the smallest (with respect to set inclusion) subset of $G(S)$ determining the gaps of S. Two different elements of $FG(S)$ are not comparable with respect to the divisibility relation.

Proposition 4.57. *Let X be a finite subset of $\mathbb{N} \setminus \{0\}$. The following conditions are equivalent.*

1) There exists a numerical semigroup S such that $FG(S) = X$.
2) $\mathbb{N} \setminus D(X)$ is a numerical semigroup and $x \nmid y$ for all $x, y \in X$ such that $x \neq y$.

Proof. 1) implies 2). This implication has been already proved.
    2) *implies* 1). If $S = \mathbb{N} \setminus D(X)$ is a numerical semigroup, then X determines its gaps. Moreover, by applying Corollary 4.56 we get that $FG(S) \subseteq X$. From the hypothesis $x \nmid y$ for all $x, y \in X$, $x \neq y$, it follows that for every $x \in X$, we have that $\{2x, 3x\} \cap D(X)$ is empty. Hence $x \in G(S)$ and $\{2x, 3x\} \subset S$. This means that $x \in FG(S)$.   □

Let S be a numerical semigroup and let $x \in SG(S)$. Then $3x = x + 2x \in S$, because $2x \in S$ by definition. Hence $SG(S) \subseteq FG(S)$. Moreover, the condition $x + s \in S$ for all $s \in S^*$ implies that the elements of $SG(S)$ are those maximal in $FG(S)$ with respect to the ordering $\leq_S$.

Proposition 4.58. *Let S be a numerical semigroup. Then*

$$SG(S) = \text{Maximals}_{\leq_S} FG(S).$$

Corollary 4.38 can be reformulated according to this information.

Corollary 4.59. *Let S be a numerical semigroup. Then S is irreducible if and only if the set $\text{Maximals}_{\leq_S} FG(S)$ has at most one element.*

The set of fundamental gaps of a numerical semigroup give an alternative way to construct the set of all its oversemigroups (see [91]).

Exercises

Exercise 4.1 ([73]). Prove that S is a pseudo-symmetric numerical semigroup with multiplicity four if and only if $S = \langle 4, x+2, x+4 \rangle$ with x an odd integer greater than or equal to three.

Exercise 4.2 ([73]). Let S be an irreducible numerical semigroup and let $n \in S$ be greater than or equal to four. If $\mathrm{Ap}(S, n) = \{0 = w_0 < w_1 < \cdots < w_{n-1}\}$, then $S' = \langle n, w_1 + n, \ldots, w_{n-2} + n \rangle$ is a numerical semigroup with multiplicity n and embedding dimension $n - 1$.

Exercise 4.3 ([73]). Let S be an irreducible numerical semigroup with $\mathrm{m}(S) \geq 5$ and $\mathrm{e}(S) = \mathrm{m}(S) - 1$. Assume that $\{n_1 < n_2 < \cdots < n_{n_1-1}\}$ is a minimal system of generators of S. Prove that $S' = \langle n_1, n_2 - n_1, \ldots, n_{n_1-1} - n_1 \rangle$ is an irreducible numerical semigroup.

Exercise 4.4 ([73]). Prove that there exists a one-to-one correspondence between the set of irreducible numerical semigroups with multiplicity $m \geq 5$ and Frobenius number F, and the set of irreducible numerical semigroups S with $\mathrm{m}(S) = m$, $\mathrm{F}(S) = F + 2m$ and with any minimal generator other than the multiplicity greater than twice the multiplicity.

Exercise 4.5 ([32, 33]). Let a and b be integers such that $0 < a < b$ and let $S = \langle a, a+1, \ldots, a+b \rangle$. Prove that S is symmetric if and only if $a \equiv 2 \bmod b$.

Exercise 4.6 ([70]). Let S be a numerical semigroup and let $x \in \mathrm{G}(S)$. Prove that there exist an irreducible numerical semigroup T with $S \subset T$ and $\mathrm{F}(T) = x$.

Exercise 4.7 ([70]). Let S be a numerical semigroup and let $\{f_1, \ldots, f_r\} = \{f \in \mathrm{PF}(S) \mid f > \mathrm{F}(S)/2\}$. Prove that there exist irreducible numerical semigroups $S_1, \ldots, S_r$ such that $\mathrm{F}(S_i) = f_i$ for all $i \in \{1, \ldots, r\}$ and $S = S_1 \cap \cdots \cap S_r$.

Exercise 4.8 ([71]). Prove that S can be expressed as an intersection of symmetric numerical semigroups if and only if for all even integer $x \in \mathrm{G}(S)$ there exists a positive odd integer y such that $x + y \notin \langle S, y \rangle$. Show that $\langle 4, 5, 6, 7 \rangle$ cannot be expressed as an intersection of symmetric numerical semigroups.

Exercise 4.9 ([71]). Show that if all the pseudo-Frobenius numbers of a numerical semigroup S are odd, then S can be expressed as an intersection of symmetric numerical semigroups. Prove that $S = \langle 5, 21, 24, 28, 32 \rangle$ is the intersection of some symmetric numerical semigroups, and that $16 \in \mathrm{PF}(S)$.

Exercise 4.10 ([91]). For a positive integer, define $D(a)$ as $D(\{a\})$. Prove that $\mathbb{N} \setminus D(a)$ is a numerical semigroup if and only if $a \in \{1, 2, 3, 4, 6\}$.

Exercise 4.11 ([91]). Let S be a numerical semigroup. Show that

$$\left\lceil \frac{\mathrm{F}(S)}{6} \right\rceil \leq \#\mathrm{FG}(S) \leq \left\lceil \frac{\mathrm{F}(S)}{2} \right\rceil.$$

Exercise 4.12 ([68]). Let S be a symmetric numerical semigroup with $m(S) \geq 3$ and let $T = S \cup \{F(S)\}$. Prove that $t(T) = e(S) = e(T) - 1$.

Exercise 4.13 ([68]). Let m and t be integers such that $1 \leq t \leq m - 1$. Show that there exists a numerical semigroup S with $t(S) = t$ and $m(S) = m$.

Exercise 4.14 ([68]). Let S be a symmetric numerical semigroup and let x be a minimal generator of S. Show that if $x < F(S)$, then

$$\{F(S), x\} \subseteq PF(S \setminus \{x\}) \subseteq \{F(S), x, F(S) - x\}.$$

Exercise 4.15 ([68]). Let m be a positive integer greater than or equal to 3. Prove that $\langle m, m + 1, \ldots, m + m - 2 \rangle$ is a symmetric numerical semigroup with Frobenius number $2m - 1$.

Exercise 4.16 ([68]). Prove that if e is an integer greater than or equal to 3, then there exists a numerical semigroup with $e(S) = e$ and $t(S) = 2$ (*Hint:* Use Exercise 4.15 with $m = e + 2$ and remove the minimal generator $2e + 2$).

Exercise 4.17 ([68]). Prove that if e is an integer greater than or equal to 4, then there exists a numerical semigroup with $e(S) = e$ and $t(S) = 3$ (*Hint:* Use Exercise 4.15 with $m = e$ and remove the minimal generator e).

Exercise 4.18 ([65]). Let S be a pseudo-symmetric numerical semigroup and let x be a minimal generator of S. Show that if $x < F(S)$, then

$$\{F(S), x\} \subseteq PF(S \setminus \{x\}) \subseteq \{F(S), x, \frac{F(S)}{2}, F(S) - x\}.$$

Exercise 4.19 ([65]). Prove that if e is an integer greater than or equal to 4, then there exists a numerical semigroup with $e(S) = e$ and $t(S) = 4$.

Exercise 4.20. Let S be an irreducible numerical semigroup and let n be a minimal generator of S. Prove that if $F(S) - n \in SG(S \setminus \{n\})$, then $(S \setminus \{n\}) \cup \{F(S) - n\}$ is also an irreducible numerical semigroup. (This gives a procedure to construct the set of all irreducible numerical semigroups with given Frobenius number.)

Exercise 4.21 ([5]). Let S be a numerical semigroup. Show that S is symmetric if and only if for every relative ideal I of S, $I^{\bullet\bullet} = I$ (see Exercise 2.13). The reader can check that the result is also true for principal relative ideals of the form $x + S$ with x a positive integer.

Exercise 4.22 ([4]). Let S be a numerical semigroup and let Ω be its canonical ideal (see Exercise 2.14). Prove that S is symmetric if and only if $S = \Omega$.

Exercise 4.23 ([34]). Let f be a positive integer. Assume that $2 \mid f$, $3 \mid f$ and $4 \mid f$. Let α and q be such that $f = 3^\alpha q$, with $\gcd\{3, q\} = 1$ (and thus $4 \mid q$). Then

$$S = \langle 3^{\alpha+1}, \frac{q}{2} + 3, 3^\alpha \frac{q}{4} + \frac{q}{2} + 3, 3^\alpha \frac{q}{2} + \frac{q}{2} + 3 \rangle$$

is an irreducible numerical semigroup with $F(S) = f$.

Exercise 4.24 ([34]). Every positive integer is the Frobenius number of an irreducible numerical semigroup with at most four generators (*Hint:* Let f be such an integer; if f is odd, use $\langle 2, f+2 \rangle$; if f is even and not a multiple of three use Lemma 4.27; if f is a multiple of two and three, and not a multiple of four, use Exercise 4.1; finally for the rest of the cases use Exercise 4.23).

Exercise 4.25 ([8]). Prove that every irreducible numerical semigroup that is not a half-line is acute (see Exercise 2.8).

Exercise 4.26 ([4]). Let S be a numerical semigroup, let M be its maximal ideal (Exercise 2.13), and let Ω be its canonical ideal (Exercise 2.14). Prove that $\#(\Omega \setminus (\Omega + M)) = \#(\Omega \setminus S) + 1$ if and only if $M = \Omega + M$. A numerical semigroup fulfilling any of these equivalent conditions is called *almost symmetric*.

Chapter 4
Proportionally modular numerical semigroups

Introduction

In [94] the authors introduce the concept of a modular Diophantine inequality. The set of integer solutions of such an inequality is a numerical semigroup. In that manuscript it is shown that the genus of these semigroups can be obtained from the coefficients of the inequality. However, to date we still do not know formulas for the Frobenius number or the multiplicity of the semigroup of solutions of a modular Diophantine inequality.

Later in [92] these inequalities are slightly modified obtaining a wider class of numerical semigroups. The new inequalities are called proportionally modular Diophantine inequalities. In [95] the concept of Bézout sequence is introduced, which became an important tool for the study of this type of numerical semigroup. These sequences are tightly related to Farey sequences (see [40] for the definition and properties of Farey sequences) and to the Stern-Brocot tree (see [38]).

1 Periodic subadditive functions

We introduce the concept of periodic subadditive function. We show that to every such mapping there exists a numerical semigroup. This correspondence also goes in the other direction; for every numerical semigroup and every nonzero element in it, we find a periodic subadditive function associated to them. The contents of this section can be found in [66].

Let $\mathbb{Q}_0^+$ denote the set of nonnegative rational numbers. A subadditive function is a map $f : \mathbb{N} \to \mathbb{Q}_0^+$ such that

(1) $f(0) = 0$,
(2) $f(x+y) \le f(x) + f(y)$ for all $x, y \in \mathbb{N}$.

From this definition it is easy to prove our next result, in which we see that every subadditive function has a submonoid of $\mathbb{N}$ associated to it.

J.C. Rosales, P.A. García-Sánchez, *Numerical Semigroups,*
Developments in Mathematics 20, DOI 10.1007/978-1-4419-0160-6_5,
© Springer Science+Business Media, LLC 2009

Lemma 5.1. *Let* $f : \mathbb{N} \to \mathbb{Q}_0^+$ *be a subadditive function. Then*

$$M(f) = \{x \in \mathbb{N} \mid f(x) \leq x\}$$

is a submonoid of $\mathbb{N}$.

Let m be a positive integer. The map $f : \mathbb{N} \to \mathbb{Q}_0^+$ has *period* m if $f(x+m) = f(x)$ for all $x \in \mathbb{N}$. We denote by $\mathscr{SF}_m$ the set of m-periodic subadditive functions. If $f \in \mathscr{SF}_m$, then we know that $M(f)$ is a submonoid of $\mathbb{N}$. Clearly, for every $x \in \mathbb{N}$ such that $x \geq \max\{f(0), \ldots, f(m-1)\}$ one has that $x \in M(f)$, which implies that $\mathbb{N} \setminus M(f)$ is finite. This proves the following lemma.

Lemma 5.2. *Let* m *be a positive integer and let* $f \in \mathscr{SF}_m$. *Then* $M(f)$ *is a numerical semigroup.*

The use of subadditive functions is inspired in the following result, which is a direct consequence of Lemma 2.6 and Proposition 3.5.

Lemma 5.3. *Let* S *be a numerical semigroup and let* m *be a nonzero element of* S. *Assume that* $\mathrm{Ap}(S,m) = \{w(0) = 0, w(1), \ldots, w(m-1)\}$ *with* $w(i) \equiv i \bmod m$ *for all* $i \in \{0, \ldots, m-1\}$. *Define* $f : \mathbb{N} \to \mathbb{N}$ *by* $f(x) = w(x \bmod m)$. *Then* $f \in \mathscr{SF}_m$ *and* $M(f) = S$.

If m is a positive integer and $f \in \mathscr{SF}_m$, then as $0 = f(0) = f(0+m) = f(m)$, we have that $f(m) \leq m$, or equivalently $m \in M(f)$, as expected.

Lemma 5.4. *Let* m *be a positive integer and* $f \in \mathscr{SF}_m$. *Then* $m \in M(f)$.

Let $\mathscr{S}_m$ be the set of numerical semigroups containing m. As a consequence of the results given so far in this section, we obtain the following result which shows the tight connection between numerical semigroups and periodic subadditive functions.

Theorem 5.5. *Let* m *be a positive integer. Then*

$$\mathscr{S}_m = \{M(f) \mid f \in \mathscr{SF}_m\}.$$

We now introduce a family of periodic subadditive functions whose associated semigroups will be the subject of study for the rest of this chapter.

Let a, b and c be positive integers. The map

$$f : \mathbb{N} \to \mathbb{Q}_0^+, \; f(x) = \frac{ax \bmod b}{c}$$

is a subadditive function of period b. Hence

$$S(a,b,c) = M(f) = \left\{x \in \mathbb{N} \;\middle|\; \frac{ax \bmod b}{c} \leq x\right\} = \{x \in \mathbb{N} \mid ax \bmod b \leq cx\}$$

is a numerical semigroup.

A *proportionally modular Diophantine inequality* is an expression of the form $ax \bmod b \leq cx$, with a, b and c positive integers. The integers a, b and c are called the *factor*, *modulus* and *proportion*, respectively. The semigroup $S(a,b,c)$ is the set of integer solutions of a proportionally modular Diophantine inequality. A numerical semigroup of this form will be called *proportionally modular*.

Example 5.6. $S(12,32,3) = \{x \in \mathbb{N} \mid 12x \bmod 32 \leq 3x\} = \{0,3,6,\rightarrow\} = \langle 3,7,8 \rangle$.

2 The numerical semigroup associated to an interval of rational numbers

We observe in this section that proportionally modular numerical semigroups are precisely the set of numerators of the fractions belonging to a bounded interval. The results of this section also appear in [92].

Given a subset A of $\mathbb{Q}_0^+$, we denote by $\langle A \rangle$, the submonoid of $\mathbb{Q}_0^+$ generated by A, that is,

$$\langle A \rangle = \{\lambda_1 a_1 + \cdots + \lambda_n a_n \mid a_1,\ldots,a_n \in A \text{ and } \lambda_1,\ldots,\lambda_n \in \mathbb{N}\}.$$

Clearly $S(A) = \langle A \rangle \cap \mathbb{N}$ is a submonoid of $\mathbb{N}$ (we use the same letter we are using for proportionally modular numerical semigroups by reasons that will become obvious later). We say that $S(A)$ is the numerical semigroup *associated* to A.

Given two rational numbers $\lambda < \mu$, we use $[\lambda,\mu]$, $[\lambda,\mu[$, $]\lambda,\mu]$ and $]\lambda,\mu[$ to denote the closed, right-opened, left-opened and opened intervals of rational numbers between λ and μ.

In this section, I denotes any of these intervals with $0 \leq \lambda < \mu$.

Lemma 5.7. *Let* $x_1,\ldots,x_k \in I$, *then* $\frac{1}{k}(x_1 + \cdots + x_k) \in I$.

Proof. As $k(\min\{x_1,\ldots,x_k\}) \leq x_1 + \cdots + x_k \leq k(\max\{x_1,\ldots,x_k\})$, we have that $\min\{x_1,\ldots,x_k\} \leq \frac{x_1+\cdots+x_k}{k} \leq \max\{x_1,\ldots,x_k\}$, and thus $\frac{1}{k}(x_1 + \cdots + x_k)$ is in I.   □

The set $S(I)$ coincides with the set of numerators of the fractions belonging to I. This fact follows from the next result.

Lemma 5.8. *Let* x *be a positive rational number. Then* $x \in \langle I \rangle$ *if and only if there exists a positive integer* k *such that* $\frac{x}{k} \in I$.

Proof. If $x \in \langle I \rangle$, then by definition $x = \lambda_1 x_1 + \cdots + \lambda_k x_k$ for some $\lambda_1,\ldots,\lambda_k \in \mathbb{N}$ and $x_1,\ldots,x_k \in I$. By Lemma 5.7, $\frac{x}{\lambda_1+\cdots+\lambda_k} \in I$.

If $\frac{x}{k} \in I$, then trivially $k\frac{x}{k} \in \langle I \rangle$.   □

We now see that every proportionally modular numerical semigroup can be realized as the numerical semigroup associated to a closed interval whose ends are determined by the factor, modulus and proportion of the semigroup.

Lemma 5.9. *Let a, b and c be positive integers with c < a. Then*

$$S(a,b,c) = S\left(\left[\frac{b}{a}, \frac{b}{a-c}\right]\right).$$

Proof. Let $x \in S(a,b,c) \setminus \{0\}$. Then $ax \bmod b \leq cx$. Hence there exists a nonnegative integer k such that $0 \leq ax - kb \leq cx$. If $k = 0$, then $ax \leq cx$, contradicting $c < a$. Thus $k \neq 0$ and $\frac{b}{a} \leq \frac{x}{k} \leq \frac{b}{a-c}$. By Lemma 5.8, we obtain $x \in S\left(\left[\frac{b}{a}, \frac{b}{a-c}\right]\right)$.

Now take $x \in S\left(\left[\frac{b}{a}, \frac{b}{a-c}\right]\right) \setminus \{0\}$. By Lemma 5.8 again, there exists a positive integer k such that $\frac{b}{a} \leq \frac{x}{k} \leq \frac{b}{a-c}$. This implies that $0 \leq ax - kb \leq cx$, and consequently $ax \bmod b \leq cx$. □

Remark 5.10. The condition $c < a$ might seem restrictive. However this is not the case, because if $c \geq a$, then the semigroup $S(a,b,c)$ is equal to $\mathbb{N}$.

Note also that the inequality $ax \bmod b \leq cx$ has the same set of integer solutions as $(a \bmod b)x \bmod b \leq cx$. Hence we can, in our study of Diophantine proportionally modular inequalities, assume that $0 < c < a < b$.

Example 5.11. $S(44,32,3) = S(12,32,3) = S\left(\left[\frac{32}{12}, \frac{32}{9}\right]\right) = S\left(\left[\frac{8}{3}, \frac{32}{9}\right]\right) = \mathbb{N} \cap (\{0\} \cup \left[\frac{8}{3}, \frac{32}{9}\right] \cup \left[\frac{16}{3}, \frac{64}{9}\right] \cup \left[8, \frac{32}{3}\right] \cup \cdots) = \{0,3,6,\rightarrow\}$.

Numerical semigroups associated to closed intervals are always proportionally modular. Its factor, modulus and proportion are determined by the ends of the interval. This result is a sort of converse to Lemma 5.9.

Lemma 5.12. *Let a_1, a_2, b_1 and b_2 be positive integers with $\frac{b_1}{a_1} < \frac{b_2}{a_2}$. Then*

$$S\left(\left[\frac{b_1}{a_1}, \frac{b_2}{a_2}\right]\right) = S(a_1 b_2, b_1 b_2, a_1 b_2 - a_2 b_1).$$

Proof. Note that $S\left(\left[\frac{b_1}{a_1}, \frac{b_2}{a_2}\right]\right) = S\left(\left[\frac{b_1 b_2}{a_1 b_2}, \frac{b_1 b_2}{b_1 a_2}\right]\right)$. The proof now follows by Lemma 5.9. □

With this we can show that the numerical semigroup associated to a bounded interval is proportionally modular.

Lemma 5.13. $S(I)$ *is a proportionally modular numerical semigroup.*

Proof. As $S(I) = \langle I \rangle \cap \mathbb{N}$, we have that $S(I)$ is a submonoid of $\mathbb{N}$. Take α and β in I with $\alpha < \beta$. Then $S([\alpha,\beta]) \subseteq S(I)$ because $[\alpha,\beta] \subseteq I$. By Lemma 5.12 and Theorem 5.5, we know that $S([\alpha,\beta])$ is a numerical semigroup, and thus has finite complement in $\mathbb{N}$. This forces $S(I)$ to have finite complement in $\mathbb{N}$, which proves that it is a numerical semigroup.

Let $\{n_1,\ldots,n_p\}$ be the minimal generating system of $S(I)$. By Lemma 5.8, there exist positive integers $d_1,\ldots,d_p$ such that $\frac{n_i}{d_i} \in I$ for all $i \in \{1,\ldots,p\}$. After rearranging the set $\{n_1,\ldots,n_p\}$, assume that

$$\frac{n_1}{d_1} < \cdots < \frac{n_p}{d_p}.$$

Then $S\left(\left[\frac{n_1}{d_1}, \frac{n_p}{d_p}\right]\right) \subseteq S(I)$, and by Lemma 5.8 again, $\{n_1, \ldots, n_p\} \subseteq S\left(\left[\frac{n_1}{d_1}, \frac{n_p}{d_p}\right]\right)$.
Thus $S\left(\left[\frac{n_1}{d_1}, \frac{n_p}{d_p}\right]\right) = S(I)$. In view of Lemma 5.12, $S(I)$ is proportionally modular. $\qquad \square$

With all these results we obtain the following characterization for proportionally modular numerical semigroups, which states that the set of solutions of a proportionally modular Diophantine inequality coincides with the set of numerators of all the fractions in a bounded interval.

Theorem 5.14. *Let S be a numerical semigroup. The following conditions are equivalent.*

1) S is proportionally modular.
2) There exist rational numbers α and β, with $0 < \alpha < \beta$, such that $S = S([\alpha, \beta])$.
3) There exists a bounded interval of positive rational numbers such that $S = S(I)$.

3 Bézout sequences

In this section we introduce the concept of Bézout sequence. As we have mentioned at the beginning of this chapter, this is one of the main tools used for the study of the set of integer solutions of a proportionally modular Diophantine inequality. This sequences and their relation with proportionally modular numerical semigroups are the main topic of [95].

A sequence of fractions $\frac{a_1}{b_1} < \frac{a_2}{b_2} < \cdots < \frac{a_p}{b_p}$ is a *Bézout sequence* if $a_1, \ldots, a_p$, $b_1, \ldots, b_p$ are positive integers such that $a_{i+1}b_i - a_ib_{i+1} = 1$ for all $i \in \{1, \ldots, p-1\}$. We say that p is the *length* of the sequence, and that $\frac{a_1}{b_1}$ and $\frac{a_p}{b_p}$ are its *ends*.

Bézout sequences are tightly connected with proportionally modular numerical semigroups. The first motivation to introduce this concept is the following property.

Proposition 5.15. *Let a_1, b_1, a_2 and b_2 be positive integers such that $a_1b_2 - a_2b_1 = 1$. Then $S\left(\left[\frac{b_1}{a_1}, \frac{b_2}{a_2}\right]\right) = \langle b_1, b_2 \rangle$.*

Proof. Let $x \in \langle b_1, b_2 \rangle \setminus \{0\}$. Then $x = \lambda b_1 + \mu b_2$ for some $\lambda, \mu \in \mathbb{N}$, not both equal to zero. As

$$\frac{b_1}{a_1} \leq \frac{\lambda b_1 + \mu b_2}{\lambda a_1 + \mu a_2} = \frac{x}{\lambda a_1 + \mu a_2} \leq \frac{b_2}{a_2},$$

in view of Lemma 5.8, $x \in S\left(\left[\frac{b_1}{a_1}, \frac{b_2}{a_2}\right]\right)$.
From Lemma 5.12, by using that $a_1b_2 - a_2b_1 = 1$, we know that

$$S\left(\left[\frac{b_1}{a_1}, \frac{b_2}{a_2}\right]\right) = S(a_1b_2, b_1b_2, 1).$$

If $x \in S\left(\left[\frac{b_1}{a_1}, \frac{b_2}{a_2}\right]\right)$, then $a_1 b_2 x \bmod b_1 b_2 \leq x$, and thus $b_2(a_1 x \bmod b_1) \leq x$. Since

$$x = \frac{x - (a_1 x \bmod b_1) b_2}{b_1} b_1 + (a_1 x \bmod b_1) b_2,$$

for proving that $x \in \langle b_1, b_2 \rangle$, it suffices to show that $\frac{x - (a_1 x \bmod b_1) b_2}{b_1} \in \mathbb{Z}$ (we already know that it is nonnegative). Or equivalently, that $(a_1 x \bmod b_1) b_2$ and x are congruent modulo b_1. Note that $(a_1 x \bmod b_1) b_2 = a_1 b_2 x \bmod b_1 b_2 = (1 + b_1 a_2) x \bmod b_1 b_2 = x + b_1 a_2 x + k b_1 b_2 = x + b_1 (a_2 x + k b_2)$ for some integer k. $\square$

Remark 5.16. Assume now that $\frac{a_1}{b_1} < \frac{a_2}{b_2} < \cdots < \frac{a_p}{b_p}$ is a Bézout sequence. From Lemma 5.8 a positive integer belongs to $S\left(\left[\frac{a_1}{b_1}, \frac{a_p}{b_p}\right]\right)$ if and only if there exists a positive integer k such that $\frac{x}{k} \in \left[\frac{a_1}{b_1}, \frac{a_p}{b_p}\right]$. Note that $\frac{x}{k} \in \left[\frac{a_1}{b_1}, \frac{a_p}{b_p}\right]$ if and only if $\frac{x}{k} \in \left[\frac{a_i}{b_i}, \frac{a_{i+1}}{b_{i+1}}\right]$ for some $i \in \{1, \ldots, p-1\}$. This is equivalent to $x \in S\left(\left[\frac{a_i}{b_i}, \frac{a_{i+1}}{b_{i+1}}\right]\right)$ in view of Lemma 5.8 again. Proposition 5.15 states then that $x \in S\left(\left[\frac{a_1}{b_1}, \frac{a_p}{b_p}\right]\right)$ if and only if $x \in \langle a_i, a_{i+1} \rangle$ for some $i \in \{1, \ldots, p-1\}$. That is,

$$S\left(\left[\frac{a_1}{b_1}, \frac{a_p}{b_p}\right]\right) = \langle a_1, a_2 \rangle \cup \langle a_2, a_3 \rangle \cup \cdots \cup \langle a_{p-1}, a_p \rangle.$$

This also proves the following.

Corollary 5.17. *Let $\frac{a_1}{b_1} < \frac{a_2}{b_2} < \cdots < \frac{a_p}{b_p}$ be a Bézout sequence. Then*

$$S\left(\left[\frac{a_1}{b_1}, \frac{a_p}{b_p}\right]\right) = \langle a_1, a_2, \ldots, a_p \rangle.$$

Example 5.18. Let us find the integer solutions to $50x \bmod 131 \leq 3x$. We know that the set of solutions to this inequality is $S\left(\left[\frac{131}{50}, \frac{131}{47}\right]\right)$. As

$$\frac{131}{50} < \frac{76}{29} < \frac{21}{8} < \frac{8}{3} < \frac{11}{4} < \frac{25}{9} < \frac{39}{14} < \frac{131}{47}$$

is a Bézout sequence, we have that $S\left(\left[\frac{131}{50}, \frac{131}{47}\right]\right) = \langle 131, 76, 21, 8, 11, 25, 39 \rangle = \langle 8, 11, 21, 25, 39 \rangle$.

In this example we have given the Bézout sequence connecting the ends of the interval defining the semigroup of solutions to the Diophantine inequality. We will soon learn how to construct such a sequence once we know the ends of an interval.

As another consequence of Proposition 5.15, we obtain that every numerical semigroup with embedding dimension two is proportionally modular.

Corollary 5.19. *Every numerical semigroup of embedding dimension two is proportionally modular.*

Proof. Let S be a numerical semigroup of embedding dimension two. There exist two relatively prime integers a and b greater than one such that $S = \langle a, b \rangle$. By Bézout's identity, there exist positive integers u and v such that $bu - av = 1$. Proposition 5.15 ensures that $S = \langle a, b \rangle = \mathrm{S}\left(\left[\frac{a}{u}, \frac{b}{v}\right]\right)$. Theorem 5.14 tells us that S is proportionally modular. $\qquad\square$

Next we will show that given two positive rational numbers, there exists a Bézout sequence whose ends are these numbers. First, we see that the numerators and denominators of the fractions belonging to an interval whose ends are rational numbers admit special expressions in terms of the numerators and denominators of these ends.

Lemma 5.20. *Let a_1, a_2, b_1, b_2, x and y be positive integers such that $\frac{a_1}{b_1} < \frac{a_2}{b_2}$. Then $\frac{a_1}{b_1} < \frac{x}{y} < \frac{a_2}{b_2}$ if and only if $\frac{x}{y} = \frac{\lambda a_1 + \mu a_2}{\lambda b_1 + \mu b_2}$ for some λ and μ positive integers.*

Proof. Necessity. If $\frac{a_1}{b_1} < \frac{x}{y} < \frac{a_2}{b_2}$, then it is not difficult to show that (x, y) belongs to the positive cone spanned by (a_1, b_1) and (a_2, b_2) (that is, to the set of pairs of the form $r(a_1, b_1) + s(a_2, b_2)$ with r and s positive rational numbers). Hence there exist positive rational numbers $\frac{p_1}{q_1}$ and $\frac{p_2}{q_2}$ such that $(x, y) = \frac{p_1}{q_1}(a_1, b_1) + \frac{p_2}{q_2}(a_2, b_2)$. Thus $q_1 q_2 x = p_1 q_2 a_1 + p_2 q_1 a_2$ and $q_1 q_2 y = p_1 q_2 b_1 + p_2 q_1 b_2$, and consequently $\frac{x}{y} = \frac{q_1 q_2 x}{q_1 q_2 y} = \frac{p_1 q_2 a_1 + p_2 q_1 a_2}{p_1 q_2 b_1 + p_2 q_1 b_2}$.

Sufficiency. Follows from the fact that for any positive integers a, b, c and d, if $\frac{a}{b} < \frac{c}{d}$, then $\frac{a}{b} < \frac{a+c}{b+d} < \frac{c}{d}$ (this has already been used in Proposition 5.15). $\qquad\square$

The next result gives the basic step for constructing a Bézout sequence whose ends are two given rational numbers.

Lemma 5.21. *Let a_1, a_2, b_1 and b_2 be positive integers such that $\frac{a_1}{b_1} < \frac{a_2}{b_2}$ and $\gcd\{a_1, b_1\} = 1$. Then there exist $x, y \in \mathbb{N} \setminus \{0\}$ such that $\frac{a_1}{b_1} < \frac{x}{y} < \frac{a_2}{b_2}$ and $b_1 x - a_1 y = 1$.*

Proof. Observe that $b_1 x - a_1 y = 1$ if and only if $x = \frac{1 + a_1 y}{b_1}$. As $\gcd\{a_1, b_1\} = 1$, the equation $a_1 y \equiv -1 \bmod b_1$ has infinitely many positive solutions. Hence $\frac{x}{y} = \frac{1 + a_1 y}{b_1 y} = \frac{a_1}{b_1} + \frac{1}{b_1 y}$ fulfills the desired inequalities for y a large-enough solution to the equation $a_1 y \equiv -1 \bmod b_1$. $\qquad\sqcap$

Among all possible values arising from the preceding lemma, we fix one that will enable us to apply induction for proving Theorem 5.23. As we will see next, this choice will allow us to effectively construct a Bézout sequence with known ends.

Lemma 5.22. *Let a_1, a_2, b_1 and b_2 be positive integers such that $\frac{a_1}{b_1} < \frac{a_2}{b_2}$, $\gcd\{a_1, b_1\} = \gcd\{a_2, b_2\} = 1$ and $a_2 b_1 - a_1 b_2 = d > 1$. Then there exists $t \in \mathbb{N}$, $1 \leq t < d$ such that $\gcd\{t a_1 + a_2, t b_1 + b_2\} = d$.*

Proof. In view of Lemma 5.21, there exist $x, y \in \mathbb{N}$ such that $\frac{a_1}{b_1} < \frac{x}{y} < \frac{a_2}{b_2}$ with $b_1 x - a_1 y = 1$. Now, from Lemma 5.20, we have that $\frac{x}{y} = \frac{\lambda a_1 + \mu a_2}{\lambda b_1 + \mu b_2}$ for some $\lambda, \mu \in \mathbb{N} \setminus \{0\}$. As $b_1 x - a_1 y = 1$, we know that $\gcd\{x, y\} = 1$ and thus $x = \frac{\lambda a_1 + \mu a_2}{\gcd\{\lambda a_1 + \mu a_2, \lambda b_1 + \mu b_2\}}$ and $y = \frac{\lambda b_1 + \mu b_2}{\gcd\{\lambda a_1 + \mu a_2, \lambda b_1 + \mu b_2\}}$. By substituting these values in $b_1 x - a_1 y = 1$ we deduce that $\gcd\{\lambda a_1 + \mu a_2, \lambda b_1 + \mu b_2\} = \mu(a_2 b_1 - a_1 b_2) = \mu d$. Hence $\mu \mid \lambda a_1 + \mu a_2$ and $\mu \mid \lambda b_1 + \mu b_2$, and consequently $\mu \mid \lambda a_1$ and $\mu \mid \lambda b_1$. By using now that $\gcd\{a_1, b_1\} = 1$, we deduce that $\mu \mid \lambda$. Let $\alpha = \frac{\lambda}{\mu} \in \mathbb{N} \setminus \{0\}$. We have then that $d = \gcd\{\alpha a_1 + a_2, \alpha b_1 + b_2\}$.

Note that if $d = \gcd\{a, b\}$, then $d \mid (a - kd, b - \bar{k}d)$ for all $k, \bar{k} \in \mathbb{N}$. By applying this fact, we deduce that if $t = \alpha \bmod d$, then $d \mid \gcd\{t a_1 + a_2, t b_1 + b_2\}$. Besides, $b_1 \frac{t a_1 + a_2}{d} - a_1 \frac{t b_1 + b_2}{d} = \frac{b_1 a_2 - a_1 b_2}{d} = \frac{d}{d} = 1$. Hence $\gcd\{\frac{t a_1 + a_2}{d}, \frac{t b_1 + b_2}{d}\} = 1$ and thus $\gcd\{t a_1 + a_2, t b_1 + b_2\} = d$.

Since $t = \alpha \bmod d$, obviously $t < d$; also $t \neq 0$, because $\gcd\{a_2, b_2\} = 1 \neq d$. $\square$

Now we are ready to show that for every two positive rational numbers, we can construct a Bézout sequence connecting them.

Theorem 5.23. *Let a_1, a_2, b_1 and b_2 be positive integers such that $\frac{a_1}{b_1} < \frac{a_2}{b_2}$, $\gcd\{a_1, b_1\} = \gcd\{a_2, b_2\} = 1$ and $a_2 b_1 - a_1 b_2 = d$. Then there exists a Bézout sequence of length less than or equal to $d + 1$ with ends $\frac{a_1}{b_1}$ and $\frac{a_2}{b_2}$.*

Proof. We proceed by induction on d. For $d = 1$ the result is trivial. Now assume that the statement holds for all the integers k with $1 \leq k < d$. By Lemma 5.22, we know that there exists a positive integer t, $1 \leq t < d$ such that $\gcd\{t a_1 + a_2, t b_1 + b_2\} = d$. Let $x_1 = \frac{t a_1 + a_2}{d}$ and $y_1 = \frac{t b_1 + b_2}{d}$. Since $\frac{x_1}{y_1} = \frac{t a_1 + a_2}{t b_1 + b_2}$, Lemma 5.20 asserts that $\frac{a_1}{b_1} < \frac{x_1}{y_1} < \frac{a_2}{b_2}$. Moreover, $b_1 x_1 - a_1 y_1 = b_1 \frac{t a_1 + a_2}{d} - a_1 \frac{t b_1 + b_2}{d} = \frac{b_1 a_2 - a_1 b_2}{d} = \frac{d}{d} = 1$ and $a_2 y_1 - b_2 x_1 = a_2 \frac{t b_1 + b_2}{d} - b_2 \frac{t a_1 + a_2}{d} = \frac{t(a_2 b_1 - a_1 b_2)}{d} = \frac{t d}{d} = t < d$. By applying the induction hypothesis to $\frac{x_1}{y_1} < \frac{a_2}{b_2}$, we deduce that there exists a Bézout sequence $\frac{x_1}{y_1} < \frac{x_2}{y_2} < \cdots < \frac{x_s}{y_s} < \frac{a_2}{b_2}$ with $s \leq t$. Hence, $\frac{a_1}{b_1} < \frac{x_1}{y_1} < \frac{x_2}{y_2} < \cdots < \frac{x_s}{y_s} < \frac{a_2}{b_2}$ is a Bézout sequence of length less than or equal to $t + 2 \leq d + 1$. $\square$

Remark 5.24. The proof of Theorem 5.23 gives an algorithmic procedure to compute a Bézout sequence with known ends $\frac{a_1}{b_1}$ and $\frac{a_2}{b_2}$. Thus we have a procedure to compute a system of generators of $S\left(\left[\frac{a_1}{b_1}, \frac{a_2}{b_2}\right]\right)$. We must first compute the least positive integer t such that $\gcd\{t a_1 + a_2, t b_1 + b_2\} = d$, and then repeat the procedure with $\left(\frac{t a_1 + a_2}{d}\right) / \left(\frac{t b_1 + b_2}{d}\right) < \frac{a_2}{b_2}$.

Example 5.25 ([95]). We start with the fractions $13/3 < 6/1$. Here $d = 5$ and so there exists $t \in \{1, \ldots, 4\}$ such that $\gcd\{13t + 6, 3t + 1\} = 5$. The choice $t = 3$ fulfills the desired condition, whence we can place $\frac{3 \times 13 + 6}{3 \times 3 + 1} = 9/2$ between $13/3$ and $6/1$. Now we proceed with $9/2 < 6/1$, and obtain $d = 3$. In this setting $\gcd\{1 \times 9 + 6, 1 \times 2 + 1\} = 3$. Thus we put $\frac{9 + 6}{2 + 1} = \frac{5}{1}$ between $9/6$ and $6/1$. Finally for $5/1 < 6/1$, it holds that $d = 1$ and consequently the process stops. A Bézout sequence for the given ends is

$$\frac{13}{3} < \frac{9}{2} < \frac{5}{1} < \frac{6}{1}.$$

Observe that Bézout sequences connecting two ends are not unique, since if $\frac{a}{b} < \frac{c}{d}$ is a Bézout sequence, then so is $\frac{a}{b} < \frac{a+c}{b+d} < \frac{c}{d}$.

4 Minimal generators of a proportionally modular numerical semigroup

We have seen the connection between systems of generators of a proportionally modular numerical semigroup and Bézout sequences. In this section we will try to sharpen this connection in order to obtain the minimal system of generators of a proportionally modular numerical semigroup. We follow the steps given in [95].

A Bézout sequence $\frac{a_1}{b_1} < \frac{a_2}{b_2} < \cdots < \frac{a_p}{b_p}$ is *proper* if $a_{i+h}b_i - a_ib_{i+h} \geq 2$ for all $h \geq 2$ such that $i, i+h \in \{1,\dots,p\}$. Every Bézout sequence can be refined to a proper Bézout sequence, by just removing those terms strictly between $\frac{a_i}{b_i}$ and $\frac{a_{i+h}}{b_{i+h}}$ whenever $a_{i+h}b_i - a_ib_{i+h} = 1$.

Example 5.26. The Bézout sequence $\frac{5}{3} < \frac{12}{7} < \frac{7}{4} < \frac{9}{5}$ is not proper, and $\frac{5}{3} < \frac{7}{4} < \frac{9}{5}$ is proper.

Lemma 5.27. *Let $\frac{a}{u} < \frac{b}{v} < \frac{c}{w}$ be a Bézout sequence. Then $b = \frac{a+c}{d}$ with $d = cu - aw$.*

Proof. The proof follows easily by taking into account that $bu - av = cv - bw = 1$. □

The next result shows that the maximum of the set of numerators of a proper Bézout sequence is always reached at one of its ends.

Lemma 5.28. *Let $\frac{a_1}{b_1} < \frac{a_2}{b_2} < \cdots < \frac{a_p}{b_p}$ be a proper Bézout sequence. Then*

$$\max\{a_1, a_2, \dots, a_p\} = \max\{a_1, a_p\}.$$

Proof. We proceed by induction on p. For $p = 2$, the statement is trivially true. We assume as induction hypothesis that $\max\{a_2, \dots, a_p\} = \max\{a_2, a_p\}$. We next show that $\max\{a_1, \dots, a_p\} = \max\{a_1, a_p\}$. If $\max\{a_2, a_p\} = a_p$, then the result follows trivially. Let us assume then that $\max\{a_2, a_p\} = a_2$. If we apply Lemma 5.27 to the Bézout sequence $\frac{a_1}{b_1} < \frac{a_2}{b_2} < \frac{a_3}{b_3}$, then we obtain that $a_2 = \frac{a_1 + a_3}{a_3b_1 - a_1b_3}$, and as this Bézout sequence is proper, $a_3b_1 - a_1b_3 \geq 2$. Hence $a_2 \leq \frac{a_1+a_3}{2} \leq \frac{2\max\{a_1, a_3\}}{2}$. We distinguish two cases depending on the value of $\max\{a_1, a_3\}$.

- If $\max\{a_1, a_3\} = a_3$, then we deduce that $a_2 \leq a_3$. Since $\max\{a_2, \dots, a_p\} = a_2$, this implies that $a_2 = a_3$. By using that $\frac{a_2}{b_2} < \frac{a_3}{b_3}$ is a Bézout sequence and $a_2 = a_3$, we obtain that $a_2(b_2 - b_3) = 1$, whence $a_2 = 1$. Since $a_1 \geq 1$, we conclude that $\max\{a_1, \dots, a_p\} = a_1$.

- If $\max\{a_1, a_3\} = a_1$, then $a_2 \leq a_1$, and the proof follows easily. $\qquad\qquad\square$

As a consequence of this result, we have that the numerators of the fractions of a proper Bézout sequence are arranged in a special way.

Proposition 5.29. *Let $\frac{a_1}{b_1} < \frac{a_2}{b_2} < \cdots < \frac{a_p}{b_p}$ be a proper Bézout sequence. Then $a_1, \ldots, a_p$ is a convex sequence, that is, there exists $h \in \{1, \ldots, p\}$ such that*

$$a_1 \geq a_2 \geq \cdots \geq a_h \leq a_{h+1} \leq \cdots \leq a_p.$$

Two fractions $\frac{a_1}{b_1} < \frac{a_2}{b_2}$ are said to be *adjacent* if

$$\frac{a_2}{b_2 + 1} < \frac{a_1}{b_1}, \text{ and } b_1 = 1 \text{ or } \frac{a_2}{b_2} < \frac{a_1}{b_1 - 1}.$$

As we will see later, this is the second condition required to obtain Bézout sequences whose numerators represent minimal systems of generators.

First we show that 1 cannot be the numerator of a fraction in a Bézout sequence of length two with adjacent ends.

Lemma 5.30. *If $\frac{a_1}{b_1} < \frac{a_2}{b_2}$ is a Bézout sequence whose ends are adjacent, then $1 \notin \{a_1, a_2\}$.*

Proof. Assume that $a_1 = 1$. Then $1 = a_2 b_1 - a_1 b_2 = a_2 b_1 - b_2$. Since $\frac{a_2}{b_2 + 1} < \frac{1}{b_1}$, we have that $a_2 b_1 < b_2 + 1$, in contradiction with $a_2 b_1 = b_2 + 1$.

Suppose now that $a_2 = 1$. Observe that in this setting $b_1 \neq 1$, since otherwise $\frac{a_1}{1} < \frac{1}{b_2}$ and thus $a_1 b_2 < 1$. Hence $\frac{1}{b_2} < \frac{a_1}{b_1 - 1}$ and therefore $b_1 - 1 < a_1 b_2$. But this is impossible because $1 = a_2 b_1 - a_1 b_2 = b_1 - a_1 b_2$. $\qquad\qquad\square$

Proposition 5.31. *If $\frac{a_1}{b_1} < \frac{a_2}{b_2} < \cdots < \frac{a_p}{b_p}$ is a proper Bézout sequence whose ends are adjacent, then $\{a_1, \ldots, a_p\}$ is the minimal system of generators of the numerical semigroup $S = \langle a_1, \ldots, a_p \rangle$.*

Proof. We use induction on p. For $p = 2$, we know by Lemma 5.30 that a_1 and a_2 are integers greater than or equal to 2 with $\gcd\{a_1, a_2\} = 1$. Thus the statement is true for $p = 2$.

From Lemma 5.28, we know that $\max\{a_1, \ldots, a_p\} = \max\{a_1, a_p\}$. We distinguish two cases, depending on the value of $\max\{a_1, a_p\}$.

- Assume that $\max\{a_1, \ldots, a_p\} = a_1$. Obviously $\frac{a_2}{b_2} < \cdots < \frac{a_p}{b_p}$ is a proper Bézout sequence. We prove that its ends are adjacent. Clearly $\frac{a_p}{b_p + 1} < \frac{a_2}{b_2}$. Note also that $b_1 \neq 1$, since otherwise the inequality $\frac{a_1}{1} < \frac{a_2}{b_2}$ would imply that $a_2 > a_1$, contradicting that $a_1 = \max\{a_1, \ldots, a_p\}$. Since $a_1 b_2 < a_2 b_1$ and $a_2 \leq a_1$, we have that $a_1 b_2 - a_1 < a_2 b_1 - a_2$. Hence, if $b_2 \neq 1$, we have that $\frac{a_p}{b_p} < \frac{a_1}{b_1 - 1} < \frac{a_2}{b_2 - 1}$. This proves that $\frac{a_2}{b_2} < \cdots < \frac{a_p}{b_p}$ is a proper Bézout sequence with adjacent ends. By using the induction hypothesis, we have that $\{a_2, \ldots, a_p\}$ minimally generates $\langle a_2, \ldots, a_p \rangle$. Since $a_1 = \max\{a_1, \ldots, a_p\}$, in order to prove that $\{a_1, \ldots, a_p\}$

is a minimal system of generators of $\langle a_1, \ldots, a_p \rangle$, it suffices to show that $a_1 \notin \langle a_2, \ldots, a_p \rangle$. In view of Corollary 5.17 we know that $\langle a_2, \ldots, a_p \rangle = \mathrm{S}\left(\left[\frac{a_2}{b_2}, \frac{a_p}{b_p}\right]\right)$. Hence, if $a_1 \in \langle a_2, \ldots, a_p \rangle$, then by Lemma 5.8 there exists a positive integer y such that $\frac{a_2}{b_2} \leq \frac{a_1}{y} \leq \frac{a_p}{b_p}$. This leads to $\frac{a_1}{b_1} < \frac{a_1}{y} \leq \frac{a_p}{b_p}$ and consequently $\frac{a_1}{b_1-1} \leq \frac{a_1}{y} \leq \frac{a_p}{b_p}$, contradicting that $\frac{a_1}{b_1}$ and $\frac{a_p}{b_p}$ are adjacent.

- Assume now that $\max\{a_1, \ldots, a_p\} = a_p$. The proof follows by arguing as in the preceding case, but now using that in this setting $\frac{a_1}{b_1} < \cdots < \frac{a_{p-1}}{b_{p-1}}$ is a proper Bézout sequence with adjacent ends. $\qquad\square$

We see next that the converse to this result also holds: every proportionally modular numerical semigroup is minimally generated by the numerators of a proper Bézout sequence with adjacent ends. The key to this result is the following lemma.

Lemma 5.32. *Let S be a proportionally modular numerical semigroup other than $\mathbb{N}$. Then there exist two minimal generators n_1 and n_p of S and positive integers b_1 and b_p such that $S = \mathrm{S}\left(\left[\frac{n_1}{b_1}, \frac{n_p}{b_p}\right]\right)$. Moreover, $\frac{n_1}{b_1}$ and $\frac{n_p}{b_p}$ are adjacent.*

Proof. Let α and β be two positive rational numbers such that $\alpha < \beta$ and $S = \mathrm{S}([\alpha, \beta])$ (Theorem 5.14). By Lemma 5.8, we know that if n is a minimal generator of S then there exists a positive integer x such that $\alpha \leq \frac{n}{x} \leq \beta$. Note that $\gcd\{n, x\} = 1$, since if $\gcd\{n, x\} = d \neq 1$, then $\alpha \leq \frac{n/d}{x/d} \leq \beta$, which would mean that $\frac{n}{d}$ is in S, contradicting that n is a minimal generator of S. Let $a(n) = \max\left\{x \in \mathbb{N} \setminus \{0\} \mid \alpha \leq \frac{n}{x}\right\}$. We are assuming that $S \neq \mathbb{N}$, thus if n_i and n_j are two distinct minimal generators of S, then $\frac{n_i}{a(n_i)} \neq \frac{n_j}{a(n_j)}$, because $\gcd\{n_i, a(n_i)\} = \gcd\{n_j, a(n_j)\} = 1$, and $\frac{n_i}{a(n_i)} = \frac{n_j}{a(n_j)}$ would imply that $n_i = n_j$. Hence there exists an arrangement of the minimal generators $n_1, \ldots, n_p$ of S such that $\alpha \leq \frac{n_1}{a(n_1)} < \frac{n_2}{a(n_2)} < \cdots < \frac{n_p}{a(n_p)} \leq \beta$. For all $i \in \{1, \ldots, p-1\}$, let $b(n_i) = \min\left\{x \in \mathbb{N} \setminus \{0\} \mid \frac{n_i}{x} \leq \frac{n_p}{a(n_p)}\right\}$. Then there exists a permutation σ on the set $\{1, \ldots, p-1\}$ such that

$$\alpha \leq \frac{n_{\sigma(1)}}{b(n_{\sigma(1)})} < \frac{n_{\sigma(2)}}{b(n_{\sigma(2)})} < \cdots < \frac{n_{\sigma(p-1)}}{b(n_{\sigma(p-1)})} < \frac{n_p}{a(n_p)} \leq \beta.$$

Note that $\alpha \leq \frac{n_{\sigma(1)}}{a(n_{\sigma(1)})} \leq \frac{n_{\sigma(1)}}{b(n_{\sigma(1)})}$ since $b(n_{\sigma(1)}) < a(n_{\sigma(1)})$, and that $\frac{n_p}{a(n_p)+1} < \alpha$ due to the maximality of $a(n_p)$. Hence $\frac{n_p}{a(n_p)+1} < \frac{n_{\sigma(1)}}{b(n_{\sigma(1)})}$. Besides, it is clear from the definition of $b(n_{\sigma(1)})$ that if $b(n_{\sigma(1)}) \neq 1$, then $\frac{n_p}{a(n_p)} < \frac{n_{\sigma(1)}}{b(n_{\sigma(1)})-1}$.

In order to conclude the proof, it suffices to show that S is the numerical semigroup $T = \mathrm{S}\left(\left[\frac{n_{\sigma(1)}}{b(n_{\sigma(1)})}, \frac{n_p}{a(n_p)}\right]\right)$. Since $\left[\frac{n_{\sigma(1)}}{b(n_{\sigma(1)})}, \frac{n_p}{a(n_p)}\right] \subseteq [\alpha, \beta]$, we have that $T \subseteq S$. As $\frac{n_{\sigma(1)}}{b(n_{\sigma(1)})} < \frac{n_{\sigma(2)}}{b(n_{\sigma(2)})} < \cdots < \frac{n_{\sigma(p-1)}}{b(n_{\sigma(p-1)})} < \frac{n_p}{a(n_p)}$, by Lemma 5.8 we deduce that $\{n_1, \ldots, n_p\} \subseteq T$. Thus $S = T$. $\qquad\square$

Proposition 5.33. *Let S be a proportionally modular numerical semigroup with $e(S) = p \geq 2$. Then there exist an arrangement $n_1, \ldots, n_p$ of the set of minimal generators of S and positive integers $b_1, \ldots, b_p$ such that $\frac{n_1}{b_1} < \frac{n_2}{b_2} < \cdots < \frac{n_p}{b_p}$ is a proper Bézout sequence with adjacent ends.*

Proof. By Lemma 5.32, we know that there exists n_1 and n_p minimal generators of S and positive integers b_1 and b_p such that $S = \mathrm{S}\left(\left[\frac{n_1}{b_1}, \frac{n_p}{b_p}\right]\right)$ and the limits of this interval are adjacent.

As we pointed out in the proof of Lemma 5.32, since n_1 and n_p are minimal generators of S and in view of Lemma 5.8, it must hold that $\gcd\{n_1, b_1\} = \gcd\{n_p, b_p\} = 1$.

If we apply Theorem 5.23 to $\frac{n_1}{b_1} < \frac{n_p}{b_p}$ and refine the resulting Bézout sequence, then we obtain a proper Bézout sequence $\frac{n_1}{b_1} < \frac{x_1}{y_1} < \cdots < \frac{x_l}{y_l} < \frac{n_p}{b_p}$ whose ends are adjacent. From Proposition 5.31, we conclude that $\{n_1, x_1, \ldots, x_l, n_p\}$ is the minimal system of generators of S. $\square$

We end this section by giving an arithmetic characterization of the minimal systems of generators of a proportionally modular numerical semigroup (and thus a characterization of these semigroups). The following easy modular computations will be useful to establish this description. Given positive integers a and b with $\gcd\{a, b\} = 1$, by Bézout's identity, there exist integers u and v such that $au + bu = 1$. We denote by $a^{-1} \mod b$ the integer $u \mod b$.

Lemma 5.34. *Let n_1 and n_2 be two integers greater than or equal to two such that $\gcd\{n_1, n_2\} = 1$. Then $n_2(n_2^{-1} \mod n_1) - n_1((-n_1)^{-1} \mod n_2) = 1$.*

Proof. Since $n_2(n_2^{-1} \mod n_1) \equiv 1 \mod n_1$ and $n_2^{-1} \mod n_1 < n_1$, we have that $\frac{n_2(n_2^{-1} \mod n_1) - 1}{n_1}$ is an integer less than n_2. Besides,

$$n_2(n_2^{-1} \mod n_1) - n_1 \frac{n_2(n_2^{-1} \mod n_1) - 1}{n_1} = 1,$$

which implies that $n_1 \frac{n_2(n_2^{-1} \mod n_1) - 1}{n_1} \equiv -1(\mod n_2)$. Hence $\frac{n_2(n_2^{-1} \mod n_1) - 1}{n_1}$ equals $(-n_1)^{-1} \mod n_2$. Thus $n_2(n_2^{-1} \mod n_1) - n_1((-n_1)^{-1} \mod n_2) = 1$. $\square$

The above-mentioned characterization is stated as follows.

Theorem 5.35. *A numerical semigroup S is proportionally modular if and only if there is an arrangement $n_1, \ldots, n_p$ of its minimal generators such that the following conditions hold:*

1) $\gcd\{n_i, n_{i+1}\} = 1$ for all $i \in \{1, \ldots, p-1\}$,
2) $n_{i-1} + n_{i+1} \equiv 0 \mod n_i$ for all $i \in \{2, \ldots, p-1\}$.

Proof. Necessity. By Proposition 5.33, we know that (possibly after a rearrangement of $n_1, \ldots, n_p$) there exist positive integers $b_1, \ldots, b_p$ such that $\frac{n_1}{b_1} < \cdots < \frac{n_p}{b_p}$ is a Bézout sequence. Hence $\gcd\{n_i, n_{i+1}\} = 1$ for all $i \in \{1, \ldots, p-1\}$. In view of Lemma 5.27, we obtain that $n_i = \frac{n_{i-1} + n_{i+1}}{n_{i+1} b_{i-1} - n_{i-1} b_{i+1}}$ for all $i \in \{2, \ldots, p-1\}$ and consequently $n_{i-1} + n_{i-1} \equiv 0 \bmod n_i$ for all $i \in \{2, \ldots, p-1\}$.

Sufficiency. From Lemma 5.34 and Condition 2), it is not hard to see that

$$\frac{n_1}{n_2^{-1} \bmod n_1} < \frac{n_2}{n_3^{-1} \bmod n_2} < \cdots < \frac{n_{p-1}}{n_p^{-1} \bmod n_{p-1}} < \frac{n_p}{(-n_{p-1})^{-1} \bmod n_p}$$

is a Bézout sequence. By Corollary 5.17 and Theorem 5.14, we conclude that S is a proportionally modular numerical semigroup. □

Example 5.36. This theorem gives a criterium to check whether or not a numerical semigroup is proportionally modular. We illustrate it with some examples.

(1) The semigroup $\langle 6, 8, 11, 13 \rangle$ is not proportionally modular, since $\gcd\{6, 8\} \neq 1$.
(2) We already know that the semigroup $\langle 8, 11, 21, 25, 39 \rangle$ is proportionally modular. Let us check it again by using the last theorem. In the arrangement of the generators described in Theorem 5.35, 8 and 11 lie together (in view of Proposition 5.29, this arrangement yields a convex sequence). It does not really matter if we start with 8,11 or 11,8, since if an arrangement fits the conditions of Theorem 5.35 so does its symmetry. The next generator we must place is 21. As $21 + 11 = 32 \equiv 0 \bmod 8$ and $21 + 8 \not\equiv 0 \bmod 11$, thus 21 goes at the left of 8. Proceeding in this way with 25 and 39, we conclude that the generators arranged as 21,8,11,25,39 fulfill the conditions of Theorem 5.35.
(3) Let us see that $\langle 5, 7, 11 \rangle$ is not proportionally modular. The generators 5 and 7 must be neighbors in the sequence. Hence we start with 5,7. If we want to place 11, then we must check if $11 + 7$ is a multiple of 5 or $5 + 11$ is a multiple of 7. None of these two conditions hold, and thus there is no possible arrangement of 5,7,11 that meets the requirements of Theorem 5.35.

5 Modular numerical semigroups

Given a, b and c positive integers, we leave open the problem of finding formulas to compute, in terms of a, b and c, the Frobenius number, genus and multiplicity of $S(a, b, c)$. In this section we present the results of [94], which show that a formula for the genus of $S(a, b, 1)$ can be given in terms of a and b.

A *modular Diophantine inequality* is an expression of the form $ax \bmod b \leq x$, with a and b positive integers. A numerical semigroup is *modular* if it is the set of solutions of a modular Diophantine inequality.

Remark 5.37. 1) Every numerical semigroup of embedding dimension two is modular (see the proof of Proposition 5.15 and Corollary 5.19).

2) There are proportionally modular numerical semigroups that are not modular (for instance $\langle 3, 8, 10 \rangle$ as shown in [92, Example 26]; this is proposed as an exercise at the end of this chapter).

Easy computations are enough to prove the following two results. We write them down because we will reference them in the future.

Lemma 5.38. *Let a and b be two integers such that $0 \leq a < b$ and let $x \in \mathbb{N}$. Then*

$$a(b-x) \bmod b = \begin{cases} 0, & \text{if } ax \bmod b = 0, \\ b - (ax \bmod b), & \text{if } ax \bmod b \neq 0, \end{cases}$$

Lemma 5.39. *Let a and b be integers such that $0 \leq a < b$. Then $ax \bmod b > x$ implies that $a(b-x) \bmod b < b - x$.*

As a consequence of this we obtain the following property, which shows that the modulus of a modular numerical semigroup behaves like the Frobenius number in a symmetric numerical semigroup.

Proposition 5.40. *Let S be a modular numerical semigroup with modulus b. If $x \in \mathbb{N} \setminus S$, then $b - x \in S$.*

As every integer greater than b belongs to $S(a,b,1)$, in order to compute the genus of $S(a,b,1)$ we can focus on the interval $[0, b-1]$. Next we see when for x in this interval, both x and $b - x$ belong to $S(a,b,1)$.

Lemma 5.41. *Let $S = S(a,b,1)$ for some integers $0 \leq a < b$, and let x be an integer such that $0 \leq x \leq b - 1$. Then $x \in S$ and $b - x \in S$ if and only if $ax \bmod b \in \{0, x\}$.*

Proof. Necessity. Assume that $ax \bmod b \neq 0$. As $x \in S$, we have that $ax \bmod b \leq x$. If $ax \bmod b < x$, then by Lemma 5.38, we have that $a(b-x) \bmod b = b - (ax \bmod b) > b - x$, and consequently $b - x \notin S$, which contradicts the hypothesis. We conclude that $ax \bmod b = x$.

Sufficiency. If $ax \bmod b = 0$, then clearly $x \in S$. Moreover, by Lemma 5.38, we have that $a(b-x) \bmod b = 0$ and thus $b - x$ is an element of S.

If $ax \bmod b = x \neq 0$, then again $x \in S$, and Lemma 5.38 states that $a(b-x) \bmod b = b - (ax \bmod b) = b - x$, which implies that $b - x \in S$. $\square$

We consider both possibilities separately. Easy modular calculations characterize them.

Lemma 5.42. *Let a and b be positive integers, and let x be an integer such that $0 \leq x \leq b - 1$. Then $ax \bmod b = 0$ if and only if x is a multiple of $\frac{b}{\gcd\{a,b\}}$.*

Lemma 5.43. *Let a and b be positive integers, and let x be an integer such that $0 \leq x \leq b - 1$. Then $ax \bmod b = x$ if and only if x is a multiple of $\frac{b}{\gcd\{a-1,b\}}$.*

With this we can control the set of integers x in $[0, b-1]$ such that $x \in S(a,b,1)$ and $b - x \in S(a,b,1)$.

Lemma 5.44. *Let $S = S(a,b,1)$ for some integers a and b such that $0 < a < b$. Let $d = \gcd\{b,a\}$ and $d' = \gcd\{b,a-1\}$, and let x be an integer such that $0 \le x \le b-1$. Then $x \in S$ and $b-x \in S$ if and only if*

$$x \in X = \left\{0, \frac{b}{d'}, 2\frac{b}{d'}, \ldots (d'-1)\frac{b}{d'}, \frac{b}{d}, 2\frac{b}{d}, \ldots, (d-1)\frac{b}{d}\right\}.$$

Moreover, the cardinality of X is $d' + d - 1$.

Proof. By Lemma 5.38 we know that $x \in S$ and $b - x \in S$ if and only if $ax \bmod b \in \{0,x\}$. By using now Lemmas 5.42 and 5.43, we know that this is equivalent to $x \in X$.

Note that $\gcd\{a-1,a\} = 1$ and thus $\gcd\{d',d\} = 1$. If $sb/d' = tb/d$ for some $s,t \in \mathbb{N}$, then $sd = td'$ and since $\gcd\{d',d\} = 1$, we deduce that there exists $k \in \mathbb{N}$ such that $sd = td' = kd'd$. Hence $s = kd'$ and $t = kd$. Therefore the cardinality of X is $d' + d - 1$.   □

The number of gaps of $S(a,b,1)$ can now be easily computed as we show in the following theorem.

Theorem 5.45. *Let $S = S(a,b,1)$ for some integers a and b with $0 \le a < b$. Then*

$$g(S) = \frac{b+1-\gcd\{a,b\}-\gcd\{a-1,b\}}{2}.$$

Proof. Let d, d' and X be as in Lemma 5.44. By using Proposition 5.40 and Lemma 5.44, we deduce that for the set $Y = \{0,\ldots,b-1\} \setminus X$, the cardinality of $(Y \cap S)$ equals that of $(Y \setminus S)$. Hence the cardinality of Y is $2g(S)$. From Lemma 5.44, we deduce that $2g(S) = b - (d+d'-1)$.   □

Open Problem 5.46. How are the minimal generators of a modular numerical semigroup characterized? More precisely, which additional condition(s) must be imposed in Theorem 5.35 to obtain a characterization of modular numerical semigroups in terms of their minimal generators?

6 Opened modular numerical semigroups

In this section we characterize those proportionally numerical semigroups that are irreducible. The idea is extracted from [97].

Recall that a numerical semigroup of the form $\{0,m,\rightarrow\}$ with m a positive integer is called a *half-line*. We say that a numerical semigroup S is an *opened modular numerical semigroup* if it is either a half-line or $S = S\left(\rbrack\frac{b}{a}, \frac{b}{a-1}\lbrack\right)$ for some integers a and b with $2 \le a < b$.

Note that the half-line $\{0,m,\rightarrow\} = S([m,2m])$ and thus it is a proportionally modular numerical semigroup in view of Theorem 5.14. The semigroups of the form $S\left(\rbrack\frac{b}{a}, \frac{b}{a-1}\lbrack\right)$ are also proportionally modular by Theorem 5.14.

We are going to see that every irreducible proportionally modular numerical semigroup is of this form. The idea is to compute the genus of these semigroups by using what we already know for modular numerical semigroups. As the Frobenius number for opened modular numerical semigroups is easy to compute, we can then search which of these semigroups have the least possible number of gaps in order to get the irreducibles.

The next result shows that opened modular numerical semigroups play the same role in the set of proportional numerical semigroups as irreducible numerical semigroups do for numerical semigroups in general.

Proposition 5.47. *Every proportionally modular numerical semigroup is the intersection of finitely many opened modular numerical semigroups.*

Proof. Let S be a proportionally modular numerical semigroup. If $S = \mathbb{N}$, then clearly S is a half-line and thus opened modular. So assume that $S \neq \mathbb{N}$. By Theorem 5.14, there exist rational numbers α and β with $1 < \alpha < \beta$ such that $S = \mathrm{S}([\alpha,\beta])$. Let $h \in \mathrm{G}(S)$. If $h \geq \alpha$, in view of Lemma 5.8, there exists $n_h \in \mathbb{N}$ such that $n_h \geq 2$ and $\frac{h}{n_h} < \alpha < \beta < \frac{h}{n_h-1}$. Define $S_h = \mathrm{S}\left(\rbrack \frac{h}{n_h}, \frac{h}{n_h-1} \lbrack \right)$, which contains S. If $h < \alpha$, set $S_h = \{0, h+1, \rightarrow\}$. Observe that in this setting $\mathrm{m}(S) > h$ (use Lemma 5.8), and consequently S_h contains S. Hence $S \subseteq \bigcap_{h \in \mathrm{G}(S)} S_h$. If $x \notin S$, then $x \in \mathrm{G}(S)$ and by Lemma 5.8 (or simply by the definition in the half-line case) $x \notin S_x$. This proves that $\bigcap_{h \in \mathrm{G}(S)} S_h \subseteq S$, and thus both semigroups coincide. $\square$

In this section, a and b represent two integers such that $2 \leq a < b$, and d and d' will denote $\gcd\{a,b\}$ and $\gcd\{a-1,b\}$, respectively.

Lemma 5.48.
$$\{b+1, \rightarrow\} \subseteq \mathrm{S}\left(\rbrack \frac{b}{a}, \frac{b}{a-1} \lbrack \right).$$

Proof. Let n be a positive integer. As $a(b+n) - (a-1)(b+n) = b+n > b$, there exists a positive integer k such that $(a-1)(b+n) < kb < a(b+n)$. This implies that $\frac{b}{a} < \frac{b+n}{k} < \frac{b}{a-1}$. Lemma 5.8 ensures that $b+n \in \mathrm{S}\left(\rbrack \frac{b}{a}, \frac{b}{a-1} \lbrack \right)$. $\square$

Lemma 5.49. *Let x be a nonnegative integer. Then*
$$x \in \mathrm{S}\left(\left[\frac{b}{a}, \frac{b}{a-1} \right] \right) \quad \text{and} \quad x \notin \mathrm{S}\left(\rbrack \frac{b}{a}, \frac{b}{a-1} \lbrack \right)$$

if and only if
$$x \in \left\{ \lambda \frac{b}{d} \,\middle|\, \lambda \in \{1,\ldots,d\} \right\} \cup \left\{ \lambda \frac{b}{d'} \,\middle|\, \lambda \in \{1,\ldots,d'\} \right\}.$$

Proof. Let $T = \mathrm{S}\left(\left[\frac{b}{a}, \frac{b}{a-1} \right] \right)$ and let $S = \mathrm{S}\left(\rbrack \frac{b}{a}, \frac{b}{a-1} \lbrack \right)$. By Lemma 5.8, if $x \in T \setminus S$, then there exists a positive integer k such that either $\frac{x}{k} = \frac{b}{a}$ or $\frac{x}{k} = \frac{b}{a-1}$. This implies that either x is a multiple of $\frac{b}{d}$ or $\frac{b}{d'}$. As by Lemma 5.48, $\{b+1, \rightarrow\} \subseteq \mathrm{S}\left(\rbrack \frac{b}{a}, \frac{b}{a-1} \lbrack \right)$, this forces $x \in \{\lambda \frac{b}{d} \mid \lambda \in \{1,\ldots,d\}\} \cup \{\lambda \frac{b}{d'} \mid \lambda \in \{1,\ldots,d'\}\}$.

For the other implication, take $x \in \{\lambda \frac{b}{d} \mid \lambda \in \{1,\dots,d\}\} \cup \{\lambda \frac{b}{d'} \mid \lambda \in \{1,\dots,d'\}\}$. Then either $x = \lambda \frac{b}{d}$ or $x = \lambda \frac{b}{d'}$. In both cases $x \in T$ by Lemma 5.8. Assume that $\lambda \frac{b}{d} \in S$. Then again by Lemma 5.8, there exists a positive integer k such that

$$ \frac{b}{a} < \frac{\lambda \frac{b}{d}}{k} < \frac{b}{a-1}. $$

And this implies that $(a-1)\lambda < dk < a\lambda$. As a is a multiple of d, both dk and $a\lambda$ are multiples of d. Since $dk < a\lambda$, we have that $dk \leq a\lambda - d$. Hence $(a-1)\lambda < a\lambda - d$, which leads to $d < \lambda$, in contradiction with the choice of λ. This proves that $\lambda \frac{b}{d} \notin S$. In a similar way it is easy to show that $\lambda \frac{b}{d'}$ is not in S. $\square$

We have achieved enough information to compute the Frobenius number and genus, with the help of Theorem 5.45, of an opened proportionally modular numerical semigroup that is not a half-line.

Theorem 5.50. *Let a and b be two integers with $2 \leq a < b$. Let $d = \gcd\{a,b\}$ and $d' = \gcd\{a-1,b\}$. Then*

$$ F\left(S\left(\left]\frac{b}{a},\frac{b}{a-1}\right[\right)\right) = b \text{ and } g\left(S\left(\left]\frac{b}{a},\frac{b}{a-1}\right[\right)\right) = \frac{1}{2}(b-1+d+d'). $$

Proof. By Lemma 5.48 and Proposition 5.49, $F\left(S\left(\left]\frac{b}{a},\frac{b}{a-1}\right[\right)\right) = b$. As $\gcd\{d,d'\} = 1$, $\lambda \frac{b}{d} \neq \lambda' \frac{b}{d'}$ for any $\lambda \in \{1,\dots,d-1\}$ and $\lambda' \in \{1,\dots,d'-1\}$. By Proposition 5.49 this implies that

$$ g\left(S\left(\left]\frac{b}{a},\frac{b}{a-1}\right[\right)\right) = g\left(S\left(\left[\frac{b}{a},\frac{b}{a-1}\right]\right)\right) + d + d' - 1. $$

We obtain the desired formula by using Theorem 5.45. $\square$

Open Problem 5.51. Even though we know formulas for the Frobenius number and genus of an opened modular numerical semigroup, a formula for the multiplicity in terms of a and b is still unknown.

From the formula given in Theorem 5.50 and the characterization of irreducible numerical semigroups established in Corollary 4.5, we get the following consequence.

Corollary 5.52. *Let a and b be integers such that $2 \leq a < b$.*

1) $S\left(\left]\frac{b}{a},\frac{b}{a-1}\right[\right)$ *is symmetric if and only if* $\gcd\{a,b\} = \gcd\{a-1,b\} = 1$.
2) $S\left(\left]\frac{b}{a},\frac{b}{a-1}\right[\right)$ *is pseudo-symmetric if and only if* $\{\gcd\{a,b\},\gcd\{a-1,b\}\} = \{1,2\}$.

Example 5.53. $S\left(\left]\frac{7}{3},\frac{7}{3-1}\right[\right) = \langle 3,5 \rangle$ is an example of the first statement. And $S\left(\left]\frac{8}{7},\frac{8}{7-1}\right[\right) = \langle 3,5,7 \rangle$ illustrates the second assertion of the last corollary.

The next result characterizes irreducible half-lines.

Lemma 5.54. *Let S be an irreducible numerical semigroup. Then S is a half-line if and only if* $S \in \{\mathbb{N}, \langle 2,3 \rangle, \langle 3,4,5 \rangle\}$.

Proof. If S is a half-line, there exists a positive integer m such that $S = \{0, m, \rightarrow\}$. Hence $S = \langle m, m+1, \ldots, 2m-1 \rangle$ and $\mathrm{e}(S) = \mathrm{m}(S)$. As S is irreducible, by Remark 4.21 and Lemma 4.15, either S has embedding dimension two or is of the form $\langle 3, x+3, 2x+3 \rangle$. Since S is a half-line, S must be either $\langle 2,3 \rangle$ or $\langle 3,4,5 \rangle$.   □

If S is not a half-line, then $\mathrm{m}(S) < \mathrm{F}(S)$. This, with the help of Lemma 5.8, translates to the following conditions in a proportionally modular numerical semigroup.

Lemma 5.55. *Let* α *and* β *be rational numbers such that* $1 < \alpha < \beta$ *and let* $S = \mathrm{S}([\alpha, \beta])$. *If S is not a half-line, then*

$$\frac{\mathrm{F}(S)}{\mathrm{F}(S)-1} < \alpha < \beta < \mathrm{F}(S).$$

Proof. As we have mentioned above, $\mathrm{m}(S) < \mathrm{F}(S)$. By Lemma 5.8, there exists a positive integer k such that $\alpha \leq \frac{\mathrm{m}(S)}{k} \leq \beta$ ($k < \mathrm{m}(S)$ because $\alpha > 1$). This leads to $\alpha \leq \frac{\mathrm{m}(S)}{k} < \frac{\mathrm{F}(S)}{k} \leq \frac{\mathrm{F}(S)}{1}$. As $\mathrm{F}(S) \notin S$, Lemma 5.8 forces $\mathrm{F}(S)$ to be greater than β. Besides, $\beta \geq \frac{\mathrm{m}(S)}{k} \geq \frac{\mathrm{m}(S)}{\mathrm{m}(S)-1} > \frac{\mathrm{F}(S)}{\mathrm{F}(S)-1}$. By using again that $\mathrm{F}(S) \notin S$ and Lemma 5.8, $\frac{\mathrm{F}(S)}{\mathrm{F}(S)-1} < \alpha$.   □

We can now prove that every irreducible proportionally modular numerical semigroup is opened modular.

Lemma 5.56. *Let S be an irreducible proportionally modular numerical semigroup that is not a half-line. Then there exists an integer k such that* $2 \leq k < \mathrm{F}(S)$ *and* $S = \mathrm{S}\left(\left]\frac{\mathrm{F}(S)}{k}, \frac{\mathrm{F}(S)}{k-1}\right[\right)$.

Proof. By Theorem 5.14, there exist rational numbers α and β such that $1 < \alpha < \beta$ and $S = \mathrm{S}([\alpha, \beta])$. From Lemmas 5.8 and 5.55 we deduce that there exists an integer k with $2 \leq k < \mathrm{F}(S)$ such that $\frac{\mathrm{F}(S)}{k} < \alpha < \beta < \frac{\mathrm{F}(S)}{k-1}$. Let $T = \mathrm{S}\left(\left]\frac{\mathrm{F}(S)}{k}, \frac{\mathrm{F}(S)}{k-1}\right[\right)$. Theorem 5.50 ensures that $\mathrm{F}(T) = \mathrm{F}(S)$. The inequalities $\frac{\mathrm{F}(S)}{k} < \alpha < \beta < \frac{\mathrm{F}(S)}{k-1}$ imply that $S \subseteq T$. The irreducibility of S forces by Theorem 4.2 that S must be equal to T, since both have the same Frobenius number.   □

With all this information, by using Corollary 4.5 it is not hard to prove the following characterization of irreducible modular numerical semigroups.

Theorem 5.57. *Let S be a proportionally modular numerical semigroup.*

1) *S is symmetric if and only if* $S = \mathbb{N}$, $S = \langle 2,3 \rangle$ *or* $S = \mathrm{S}\left(\left]\frac{b}{a}, \frac{b}{a-1}\right[\right)$ *for some integers a and b with* $2 \leq a < b$ *and* $\gcd\{a,b\} = \gcd\{a-1,b\} = 1$.
2) *S is pseudo-symmetric if and only if* $S = \langle 3,4,5 \rangle$ *or* $S = \mathrm{S}\left(\left]\frac{b}{a}, \frac{b}{a-1}\right[\right)$ *for some integers a and b with* $2 \leq a < b$ *and* $\gcd\{a,b\} = \gcd\{a-1,b\} = \{1,2\}$.

Exercises

Exercise 5.1. Let a, b and c be positive integers with $\gcd\{a,b\} = 1$. Prove that $S = \langle a, a+b, a+2b, \ldots, a+cb \rangle$ is a proportionally modular numerical semigroup.

Exercise 5.2. Let S be a proportionally modular numerical semigroup with minimal system of generators $\{n_1 < n_2 < \cdots < n_e\}$ and $e \geq 3$. Prove that $\langle n_1, \ldots, n_{e-1} \rangle$ is also a proportionally modular numerical semigroup.

Exercise 5.3 ([25]). Given integers a, b and c such that $0 < c < a < b$, prove that

$$S(a,b,c) = S(b+c-a, b, c).$$

Exercise 5.4 ([95]). Prove that a numerical semigroup S is proportionally modular if and only if there is an arrangement $n_1, \ldots, n_e$ of its minimal generators such that the following conditions hold:

1) $\langle n_i, n_{i+1} \rangle$ is a numerical semigroup for all $i \in \{1, \ldots, e-1\}$,
2) $\langle n_{i-1}, n_i, n_{i+1} \rangle = \langle n_{i-1}, n_i \rangle \cup \langle n_i, n_{i+1} \rangle$ for all $i \in \{2, \ldots, e-1\}$.

(*Hint:* Use Theorem 5.35.) Observe that this result sharpens the contents of Remark 5.16.

Exercise 5.5. Let $S = \langle 7, 8, 9, 10, 12 \rangle$. Prove that S is not proportionally modular. However $S = \langle 12, 7 \rangle \cup \langle 7, 8 \rangle \cup \langle 8, 9 \rangle \cup \langle 9, 10 \rangle$.

Exercise 5.6. Find two proportionally modular numerical semigroups whose intersection is not proportionally modular.

Exercise 5.7. Give an example of a proportionally modular numerical semigroup $S \neq \mathbb{N}$ such that $S \cup \{F(S)\}$ is not proportionally modular.

Exercise 5.8 ([25]). For integers a, b and c with $0 < c < a < b$, prove that

$$F(S(a,b,c)) = b - \left\lfloor \frac{\zeta b}{a} \right\rfloor - 1,$$

where $\zeta = \min \left\{ k \in \{1, \ldots, a-1\} \mid kb \bmod a + \left\lfloor \frac{kb}{a} \right\rfloor c > (c-1)b + a - c \right\}$.

Exercise 5.9 ([94]). Let $ax \bmod b \leq x$ be a modular Diophantine inequality (with as usual $0 < a < b$). We define its *weight* as $w(a,b) = b - \gcd\{a,b\} - \gcd\{a-1,b\}$.

a) Prove that if two modular Diophantine inequalities have the same set of integers solutions, then they have the same weight.
b) Find an example showing that the converse of a) does not hold in general.
c) Prove that $w(a,b)$ is an odd integer greater than or equal to $F(S(a,b,1))$.
d) Show that $S(a,b,1)$ is symmetric if and only if $w(a,b) = F(S(a,b,1))$.
e) Show that $S(a,b,1)$ is pseudo-symmetric if and only if $w(a,b) = F(S(a,b,1))+1$.

Exercise 5.10 ([94]). Let a and b be integers with $0 < a < b$. Prove that $b \geq F(S(a,b,1)) + m(S(a,b,1))$ and that the equality holds if and only if

$$m(S(a,b,1)) \neq \min\left\{\frac{b}{\gcd\{a,b\}}, \frac{b}{\gcd\{a-1,b\}}\right\}.$$

Exercise 5.11 ([94]). Given integers a and b with $0 < a < b$, show that

$$b \leq 12g(S(a,b,1)) - 6.$$

Exercise 5.12. Prove that $S = \langle 3, 8, 10 \rangle$ is a proportionally modular numerical semigroup that is not modular.

Exercise 5.13 ([94]). Let a and b be positive integers. Prove that

a) $m(S(a,ab,1)) = b$,
b) $F(S(a,ab,1)) = \left\lceil \frac{(a-1)(b-1)}{b} \right\rceil b - 1$.

Exercise 5.14 ([94]). Let a and b be integers such that $0 < a < b$ and $b \bmod a \neq 0$. Show that

a) $F(S(a,b,1)) = b - \left\lceil \frac{b}{a} \right\rceil$ if and only if $(a-1)(a-(b \bmod a)) < b$,
b) if $(a-1)(a-(b \bmod a)) < b$, then $m(S(a,b,1)) = \left\lceil \frac{b}{a} \right\rceil$.

Chapter 5
The quotient of a numerical semigroup by a positive integer

Introduction

A generalization of the linear Diophantine Frobenius problem can be stated as follows. Let $n_1, \ldots, n_p$ and d be positive integers with $\gcd\{n_1, \ldots, n_p\} = 1$. Find a formula for the largest multiple of d not belonging to $\langle n_1, \ldots, n_p \rangle$. This problem is equivalent to the computation of the Frobenius number of the semigroup $\frac{\langle n_1, \ldots, n_p \rangle}{d}$, and it still remains open for $p = 2$.

Semigroups of the form $\frac{S}{d}$ also occur in a natural way in the scope of proportionally modular numerical semigroups. In [98], it is shown that a numerical semigroup is proportionally modular if and only if it is the quotient of an embedding dimension two numerical semigroup. This result is later sharpened in [54] where it is shown that it suffices to take numerical semigroups generated by an integer and this integer plus one. So far we have no general formula for the largest multiple of an integer not belonging to $\langle a, a+1 \rangle$, with a an integer greater than two.

Since numerical semigroups with embedding dimension two are symmetric, we wondered which is the class of all numerical semigroups that are quotients of symmetric numerical semigroups. Surprisingly, this class covers the set of all numerical semigroups as shown in [83]. What is more amazing is that it suffices to divide by two. The same does not hold for pseudo-symmetric numerical semigroups, and we need to divide by four to obtain the whole set of numerical semigroups as quotients of pseudo-symmetric numerical semigroups (see [69]). As for other families of numerical semigroups, for instance, we still do not know how to decide if a numerical semigroup is the quotient of a numerical semigroup with embedding dimension three.

In [106] a class of numerical semigroups is presented whose elements are the positive cones of the K_0 groups of some C^*-algebras. These semigroups are intersections of quotients of embedding dimension two numerical semigroups under several extra conditions. It turns out that this class coincides with that of finite intersections of proportionally modular numerical semigroups (see [84]).

J.C. Rosales, P.A. García-Sánchez, *Numerical Semigroups*,
Developments in Mathematics 20, DOI 10.1007/978-1-4419-0160-6_6,
© Springer Science+Business Media, LLC 2009

1 Notable elements

We describe in this section those notable elements that are easy to compute in a quotient of a numerical semigroup by an integer (once these notable elements are known in the original semigroup).

Let S be a numerical semigroup and let p be a positive integer. Set

$$\frac{S}{p} = \{x \in \mathbb{N} \mid px \in S\}.$$

Proposition 6.1. *Let S be a numerical semigroup and let p be a positive integer.*

1) $\frac{S}{p}$ *is a numerical semigroup.*
2) $S \subseteq \frac{S}{p}$.
3) $\frac{S}{p} = \mathbb{N}$ *if and only if $p \in S$.*

The semigroup $\frac{S}{p}$ is called the *quotient* of S by p. Accordingly we say that $\frac{S}{2}$ is one half of S and that $\frac{S}{4}$ is one fourth of S. We mention these two particular instances because they will be of some relevance later in this chapter.

As we have seen in Section 5 of Chapter 3, the fundamental gaps of a numerical semigroup determine it uniquely. Fortunately, the fundamental gaps of the quotient of a numerical semigroup can be relatively easily calculated from the fundamental gaps of the original semigroup. This does not hold for minimal generators; this is why we focus on the fundamental gaps of the numerical semigroup.

Proposition 6.2 ([92]). *Let S be a numerical semigroup and let d be a positive integer. Then*

$$\mathrm{FG}\left(\frac{S}{d}\right) = \left\{\frac{h}{d} \;\middle|\; h \in \mathrm{FG}(S) \text{ and } h \equiv 0 \bmod d\right\}.$$

Proof. The integer h belongs to $\mathrm{FG}(\frac{S}{d})$ if and only if $h \notin \frac{S}{d}$ and $kh \in \frac{S}{d}$ for every integer k greater than one. This is equivalent to $dh \notin S$ and $kdh \in S$ for any integer k greater than one. □

Corollary 6.3. *Let S be a numerical semigroup and let d be a positive integer. Then $d \in \mathrm{FG}(S)$ if and only if $\frac{S}{d} = \langle 2,3 \rangle$.*

Proof. Observe that $\mathrm{FG}(\langle 2,3 \rangle) = 1$. Then use Proposition 6.2. □

Example 6.4.

```
gap> S:=NumericalSemigroup(5,7,8);
<Numerical semigroup with 3 generators>
gap> FrobeniusNumberOfNumericalSemigroup(S);
11
gap> List([1..11],
> d->QuotientOfNumericalSemigroup(S,d));;
```

```
gap> List(last,
> MinimalGeneratingSystemOfNumericalSemigroup);
[ [ 5, 7, 8 ], [ 4, 5, 6, 7 ], [ 4, 5, 6, 7 ],
[ 2, 3 ], [ 1 ], [ 2, 3 ], [ 1 ], [ 1 ], [ 2, 3 ],
[ 1 ], [ 2, 3 ] ]
```

As we know, one of the best ways to describe a numerical semigroup is by means of the Apéry set of any of its nonzero elements. Note that if S is a numerical semigroup, $m \in S^*$ and $d \mid m$, then $\frac{m}{d} \in \frac{S}{d}$. We describe $\mathrm{Ap}\left(\frac{S}{d}, \frac{m}{d}\right)$ in terms of $\mathrm{Ap}(S, m)$.

Proposition 6.5. *Let S be a numerical semigroup. Let m be a nonzero element of S and let d be a divisor of m. Then*

$$\mathrm{Ap}\left(\frac{S}{d}, \frac{m}{d}\right) = \left\{\frac{w}{d} \;\middle|\; w \in \mathrm{Ap}(S, m) \text{ and } w \equiv 0 \bmod d\right\}.$$

Proof. The idea of the proof is analogous to the proof of Proposition 6.2. □

By now using Selmer's formula (Proposition 2.12), we obtain the following nice consequence.

Corollary 6.6. *Let S be a numerical semigroup. Let m be a nonzero element of S and let d be a divisor of m. Assume that $\mathrm{Ap}(S, m) = \{0, k_1 m + 1, \ldots, k_{m-1} m + m - 1\}$. Then*

1) $\mathrm{Ap}\left(\frac{S}{d}, \frac{m}{d}\right) = \{0, k_d \frac{m}{d} + 1, \ldots, k_{(\frac{m}{d}-1)d} \frac{m}{d} + \frac{m}{d} - 1\}$,
2) $\mathrm{g}\left(\frac{S}{d}\right) = k_d + k_{2d} + \cdots + k_{(\frac{m}{d}-1)d}$,
3) $\mathrm{F}\left(\frac{S}{d}\right) = \max\{k_d \frac{m}{d} + 1, \ldots, k_{(\frac{m}{d}-1)d} \frac{m}{d} + \frac{m}{d} - 1\} - \frac{m}{d}$.

2 One half of an irreducible numerical semigroup

In this section we show (following [83]) that every numerical semigroup is one half of infinitely many symmetric numerical semigroups. We will also prove that every numerical semigroup is one fourth of a pseudo-symmetric numerical semigroup (by using to this end the ideas given in [69]).

For a numerical semigroup S, set

$$2S = \{2s \mid s \in S\}.$$

This set is a submonoid of $\mathbb{N}$. Moreover, $2\langle n_1, \ldots, n_p \rangle = \langle 2n_1, \ldots, 2n_p \rangle$. Recall that if A and B are subsets of integer numbers, we write $A + B = \{a + b \mid a \in A, b \in B\}$ (note that in general with this notation $A + A \neq 2A$).

Theorem 6.7. *Let $S = \langle n_1, \ldots, n_p \rangle$ with $\mathrm{PF}(S) = \{f_1, \ldots, f_t\}$. Let f be an odd integer such that $f - f_i - f_j \in S$ for all $i, j \in \{1, \ldots, t\}$. Then*

$$T = \langle 2n_1, 2n_2, \ldots, 2n_p, f - 2f_1, \ldots, f - 2f_t \rangle$$

is a symmetric numerical semigroup with Frobenius number f and $S = \frac{T}{2}$. Moreover,
$T = 2S \cup (\{f - 2f_1, \ldots, f - 2f_t\} + 2S)$.

Proof. Let $i, j \in \{1, \ldots, t\}$. Observe that $f - 2f_i + f - 2f_j = 2(f - f_i - f_j)$, which by hypothesis belongs to $2S$.

From the above remark, it easily follows that

$$T = 2S \cup (\{f - 2f_1, \ldots f - 2f_t\} + 2S), \quad T \cap 2\mathbb{N} = 2S. \tag{1}$$

We see now that $f \notin T$. If this were not the case, then as f is odd, there would exist $i \in \{1, \ldots, t\}$ and $s \in S$ such that $f = f - 2f_i + 2s$. But this would lead to $f_i = s \in S$, which is impossible.

Now we prove that all even integers greater than f are in T. As $F(S) = \max\{f_1, \ldots, f_t\}$, by hypothesis, we have that $f - F(S) - F(S) \in S$ and thus $f \geq 2F(S)$. Clearly, every positive even integer greater than $2F(S)$ is in $2S$. Hence, every even integer greater than f is in $2S \subseteq T$.

Next we show that $F(T) = f$. From the preceding paragraphs, it suffices to show that every odd integer greater than f belongs to T. Let $k \in \mathbb{N} \setminus \{0\}$. Then $f + 2k = (f - 2F(S)) + 2(F(S) + k)$. As $f - 2F(S) \in \{f - 2f_1, \ldots, f - 2f_t\}$ and $2(F(S) + k) \in 2S$, we deduce in view of (1) that $f + 2k \in T$.

In order to prove that T is symmetric, take $x \in \mathbb{Z} \setminus T$. We must show that $f - x \in T$ (Proposition 4.4). We distinguish two cases depending on the parity of x.

- If x is even, then as $x \notin T$, we have that $\frac{x}{2} \notin S$. In view of Proposition 2.19, there exists $i \in \{1, \ldots, t\}$ such that $f_i - \frac{x}{2} \in S$. Thus $2f_i - x \in 2S$. Hence from (1), we have that $f - x = (f - 2f_i) + (2f_i - x) \in T$.
- If x is odd, then $f - x$ is even. Thus if $f - x \notin T$, by using the preceding case, we obtain that $f - (f - x) = x \in T$, contradicting the choice of x.

Finally, we prove that $S = \frac{T}{2}$. If $x \in \frac{T}{2}$, then $2x \in T$. In view of (1), this means that $2x \in 2S$, whence $x \in S$. Conversely, if $x \in S$, then $2x \in 2S \subseteq T$, which leads to $x \in \frac{T}{2}$. $\square$

Next we see as a consequence of this theorem that we can choose infinitely many T for every semigroup S.

Corollary 6.8. *Let S be a numerical semigroup. Then there exist infinitely many symmetric numerical semigroups T such that $S = \frac{T}{2}$.*

Proof. Assume that $PF(S) = \{f_1, \ldots, f_t\}$. Choose an odd integer f greater than or equal to $3F(S) + 1$. For $i, j \in \{1, \ldots, t\}$, $f - f_i - f_j \geq 3F(S) + 1 - F(S) - F(S) = F(S) + 1$. Thus $f - f_i - f_j \in S$. From Theorem 6.7, we know that $T_f = 2S \cup (\{f - 2f_1, \ldots, f - 2f_t\} + 2S)$ is a symmetric numerical semigroup with Frobenius number f and such that $S = \frac{T_f}{2}$. The proof now follows by observing that we can choose infinitely many odd numbers greater than or equal to $3F(S) + 1$, and that for each of them we obtain a different T_f. $\square$

Now let us proceed with the pseudo-symmetric case. As a consequence of the following result, we will see that there is not a parallelism between the symmetric and pseudo-symmetric case.

Lemma 6.9. *Let S be a numerical semigroup with even Frobenius number. Then*

$$\mathrm{F}\left(\frac{S}{2}\right) = \frac{\mathrm{F}(S)}{2}.$$

Proof. Follows by Proposition 6.2, by taking into account that the Frobenius number of a numerical semigroup is the maximum of the fundamental gaps.   □

Remark 6.10. Note that if S is a numerical semigroup and T is a pseudo-symmetric numerical semigroup with $S = \frac{T}{2}$, then from the preceding lemma we deduce that $\mathrm{F}(T) = 2\mathrm{F}(S)$. It is clear that there exist finitely many numerical semigroups with Frobenius number $2\mathrm{F}(S)$. Hence we cannot obtain a result similar to Corollary 6.8 for the pseudo-symmetric case.

We even have another issue that makes the pseudo-symmetric case different from the symmetric case, as we see in the next proposition.

Proposition 6.11. *Let T be a pseudo-symmetric numerical semigroup. Then $\frac{T}{2}$ is an irreducible numerical semigroup.*

Proof. Let $S = \frac{T}{2}$ and suppose that $\mathrm{PF}(S) = \{f_1 < \cdots < f_t\}$. Recall that $\mathrm{F}(S)$ is the maximum of the pseudo-Frobenius numbers. Hence $\mathrm{F}(S) = f_t$. Thus, by applying Lemma 6.9, we get $\mathrm{F}(T) = 2f_t$.

Let $i \in \{1,\dots,t\}$. Then $f_i \notin S$ and so $2f_i \notin T$. From Proposition 4.4 we have that either $2f_t - 2f_i \in T$ or $2f_i = f_t$. If $2f_t - 2f_i \in T$, then $2(f_t - f_i) \in T$. Hence $f_t - f_i \in S$ and, in view of Proposition 2.19, we obtain $f_t = f_i$. This proves that $\mathrm{PF}(S) \subseteq \left\{ f_t, \frac{f_t}{2} \right\} = \left\{ \mathrm{F}(S), \frac{\mathrm{F}(S)}{2} \right\}$. In view of Corollaries 4.11 and 4.16, we have that S is either a symmetric or a pseudo-symmetric numerical semigroup. In both cases, S is an irreducible numerical semigroup.   □

This in particular means that we cannot obtain that every numerical semigroup is one half of a pseudo-symmetric numerical semigroup.

We can sharpen a bit more the preceding result in order to distinguish in which cases one half of a pseudo-symmetric numerical semigroup is symmetric or pseudo-symmetric.

Corollary 6.12. *Let S be a pseudo-symmetric numerical semigroup. Then*

1) $\frac{S}{2}$ *is symmetric if and only if* $\mathrm{F}(S) \not\equiv 0 \bmod 4$,
2) $\frac{S}{2}$ *is pseudo-symmetric if and only if* $\mathrm{F}(S) \equiv 0 \bmod 4$.

Proof. This is a consequence of Lemma 6.9, Proposition 6.11 and Proposition 4.4.   □

Now we see that every symmetric numerical semigroup is one half of a pseudo-symmetric numerical semigroup. This together with the fact that every numerical semigroup is one half of a symmetric semigroup will prove that any numerical semigroup can be expressed as one fourth of infinitely many pseudo-symmetric numerical semigroups.

Lemma 6.13. *Let S be a symmetric numerical semigroup. Let*

$$A = \left\{ F(S) + 2k \mid k \in \left\{ 1, \ldots, \frac{F(S) - 1}{2} \right\} \right\}.$$

Then

$$T = 2S \cup A \cup \{2F(S) + 1, \rightarrow\}$$

is a pseudo-symmetric numerical semigroup with Frobenius number $2F(S)$ and such that $S = \frac{T}{2}$.

Proof. Since S is symmetric, $F(S)$ is odd. Notice that A is the set of odd integers belonging to the set $\{F(S) + 2, \ldots, 2F(S) - 1\}$ and that $\#A = \frac{F(S) - 1}{2}$.

We start by proving that T is a numerical semigroup. In one hand it is obvious that the sum of two elements of $2S$ is an element of $2S$ and that the result of adding any nonnegative integer to any element in $\{2F(S) + 1, \rightarrow\}$ remains in $\{2F(S) + 1, \rightarrow\}$. On the other hand the sum of elements of A is an element of $\{2F(S) + 1, \rightarrow\}$. Finally the sum of an element of $2S$ with an element of A is an element of $A \cup \{2F(S) + 1, \rightarrow\}$. Notice also that since $\{2F(S) + 1, \rightarrow\} \subseteq T$, we have that $\mathbb{N} \backslash T$ is finite.

Now let us prove that $2F(S)$ is the Frobenius number of T. As $\{2F(S) + 1, \rightarrow\} \subseteq T$, we only have to show that $2F(S) \notin T$. But this holds since $2F(S)$ is an even number and $2F(S) \notin 2S$.

Next we will see that T is a pseudo-symmetric numerical semigroup. In view of Corollary 4.5 and since $2F(S)$ is the Frobenius number of T, it suffices to prove that $n(T) = F(S)$. As S is symmetric, by Corollary 4.5 we have $n(S) = \frac{F(S)+1}{2}$. Hence $\#\{x \in 2S \mid x \le 2F(S)\} = \#\{x \in S \mid x \le F(S)\} = n(S) = \frac{F(S)+1}{2}$. Therefore $n(T) = \frac{F(S)+1}{2} + \#A = \frac{F(S)+1}{2} + \frac{F(S)-1}{2} = F(S)$.

Finally we prove that $S = \frac{T}{2}$. We have $x \in \frac{T}{2}$ if and only if $2x \in T$. Since the elements of A are odd, we obtain $2x \in T$ if and only if $2x \in 2S \cup \{2F(S) + 1, \rightarrow\}$. If $2x \in 2S$, then trivially $x \in S$. If $2x \ge 2F(S) + 1$, then $x \ge F(S) + 1$ and thus $x \in S$. Therefore $x \in \frac{T}{2}$ if and only if $x \in S$. $\square$

As a consequence of this lemma and Corollary 6.8, we obtain the following.

Theorem 6.14. *Every numerical semigroup is one fourth of infinitely many pseudo-symmetric numerical semigroups.*

We can achieve a result similar to Lemma 6.13 for pseudo-symmetric numerical semigroups. This will help us to establish a new characterization of irreducible numerical semigroups.

Lemma 6.15. *Let S be a pseudo-symmetric numerical semigroup. Let*

$$A = \left\{ F(S) + 2k - 1 \,\middle|\, k \in \left\{ 1, \ldots, \frac{F(S)}{2} \right\} \right\}.$$

Then

$$T = 2S \cup A \cup \{2F(S) + 1, \rightarrow\}$$

is a pseudo-symmetric numerical semigroup with Frobenius number 2F(S) *and such that* $S = \frac{T}{2}$.

Proof. Since S is pseudo-symmetric, F(S) is even. Notice that A is the set of odd integers belonging to the set $\{F(S) + 1, \ldots, 2F(S) - 1\}$ and that $\#A = \frac{F(S)}{2}$.

The proof that T is a numerical semigroup with Frobenius number 2F(S) and that $S = \frac{T}{2}$ is similar to the one performed in Lemma 6.13.

Let us see that T is pseudo-symmetric. In view of Corollary 4.5 and the fact that 2F(S) is the Frobenius number of T, it suffices to prove that $n(T) = F(S)$. Since S is a pseudo-symmetric numerical semigroup, from Corollary 4.5 we deduce that $n(S) = \frac{F(S)}{2}$. Hence $\#\{x \in 2S \mid x \leq 2F(S)\} = \#\{x \in S \mid x \leq F(S)\} = n(S) = \frac{F(S)}{2}$. Therefore $n(T) = \frac{F(S)}{2} + \#A = \frac{F(S)}{2} + \frac{F(S)}{2} = F(S)$. $\square$

With all this information we get a new characterization of irreducible numerical semigroups.

Theorem 6.16. *A numerical semigroup is irreducible if and only if it is one half of a pseudo-symmetric numerical semigroup.*

3 Numerical semigroups having a Toms decomposition

Let S be a numerical semigroup. According to [106], we say that S has a *Toms decomposition* if there exist $q_1, \ldots, q_n, m_1, \ldots, m_n$ and L such that

1) $\gcd(\{q_i, m_i\}) = \gcd(\{L, q_i\}) = \gcd(\{L, m_i\}) = 1$ for all $i \in \{1, \ldots, n\}$,
2) $S = \frac{1}{L} \bigcap_{i=1}^{n} \langle q_i, m_i \rangle$.

Let a, b and c be positive integers. We say that the monoid $\frac{\langle a,b \rangle}{c}$ is a *Toms block* if $\gcd(\{a,b\}) = \gcd(\{a,c\}) = \gcd(\{b,c\}) = 1$. As we are imposing the condition $\gcd(\{a,b\}) = 1$, every Toms block is a numerical semigroup. Observe that $\frac{1}{L} \bigcap_{i=1}^{n} \langle q_i, m_i \rangle = \bigcap_{i=1}^{n} \frac{\langle q_i, m_i \rangle}{L}$. So a numerical semigroup admits a Toms decomposition if and only if it can be expressed as an intersection of finitely many Toms blocks with the same denominator.

Our aim in this section is to prove that a numerical semigroup admits a Toms decomposition if and only if it is the intersection of finitely many proportionally modular numerical semigroups. We start by showing that the set of proportionally modular numerical semigroups coincides with the set of quotients of embedding dimension two numerical semigroups. First we need to show that the numerators of the ends of an interval of rational numbers defining a numerical semigroup can be chosen to be coprime.

Lemma 6.17. *Let a, b and c be positive integers with $c < a < b$. Then there exist positive integers k and d such that* $\mathrm{S}\left(\left[\frac{b}{a}, \frac{b}{a-c}\right]\right) = \mathrm{S}\left(\left[\frac{b}{a}, \frac{kb+1}{d}\right]\right)$.

Proof. Let $S = \mathrm{S}\left(\left[\frac{b}{a}, \frac{b}{a-c}\right]\right)$. By Lemma 5.8, if $x \in \mathbb{N} \setminus S$, then there exists a unique number $n_x \in \mathbb{N}$ such that $\frac{x}{n_x+1} < \frac{b}{a} < \frac{b}{a-c} < \frac{x}{n_x}$. Since S is a numerical semigroup, then $\mathbb{N} \setminus S$ is finite and so there exists the minimum q of the set $\left\{\frac{x}{n_x} \mid x \in \mathbb{N} \setminus S\right\}$. Note that $\frac{b}{a-c} < q$ and in consequence there exist two positive integers d, k such that $d\frac{b}{a-c} \leq kb + 1 < dq$. The interval inclusion $\left[\frac{b}{a}, \frac{b}{a-c}\right] \subseteq \left[\frac{b}{a}, \frac{kb+1}{d}\right]$ implies that $S \subseteq \mathrm{S}\left(\left[\frac{b}{a}, \frac{kb+1}{d}\right]\right)$. To show that $\mathrm{S}\left(\left[\frac{b}{a}, \frac{kb+1}{d}\right]\right) \subseteq S$, take $x \in \mathrm{S}(\left[\frac{b}{a}, \frac{kb+1}{d}\right])$. By Lemma 5.8, there exists a positive integer y such that $\frac{b}{a} \leq \frac{x}{y} \leq \frac{kb+1}{d}$. Since $\frac{x}{y} < q$, this implies that $x \in S$. □

Theorem 6.18 ([98]). *Let n_1, n_2 and d be positive integers with n_1 and n_2 relatively prime. Then $\frac{\langle n_1, n_2\rangle}{d}$ is a proportionally modular numerical semigroup. Conversely, every proportionally modular numerical semigroup can be represented in this form.*

Proof. As $\gcd\{n_1, n_2\} = 1$, there exist two positive integers u, v such that $un_2 - vn_1 = 1$. By Lemma 5.16 and Proposition 5.21 we know that

$$\langle n_1, n_2\rangle = \{x \in \mathbb{N} \mid un_2x \bmod n_1 n_2 \leq x\},$$

and it is not hard to see that

$$\frac{\langle n_1, n_2\rangle}{d} = \{x \in \mathbb{N} \mid un_2 dx \bmod n_1 n_2 \leq dx\},$$

which shows that $\frac{\langle n_1, n_2\rangle}{d}$ is a proportionally modular numerical semigroup.

Suppose now that S is a proportionally modular numerical semigroup defined by the condition $ax \bmod b \leq cx$. By Lemma 5.9 we know that $S = \mathrm{S}\left(\left[\frac{b}{a}, \frac{b}{a-c}\right]\right)$. From Lemma 6.17, we deduce that there exist positive integers a_1, b_1, a_2, and b_2 such that $\frac{a_1}{b_1} < \frac{a_2}{b_2}$, $\gcd\{a_1, a_2\} = 1$ and $S = \mathrm{S}\left(\left[\frac{a_1}{b_1}, \frac{a_2}{b_2}\right]\right)$. By Lemma 5.12, we have that $S = \{x \in \mathbb{N} \mid a_2 b_1 x \bmod a_1 a_2 \leq dx\}$, with $d = a_2 b_1 - a_1 b_2$.

We show that $S = \frac{\langle a_1, a_2\rangle}{d}$. The condition $\gcd\{a_1, a_2\} = 1$ implies that there exist positive integers u, v such that $a_2 u - a_1 v = 1$. Hence we have that

$$\frac{\langle a_1, a_2\rangle}{d} = \{x \in \mathbb{N} \mid ua_2 dx \bmod a_1 a_2 \leq dx\}.$$

To conclude the proof, we check that the inequalities $a_2 b_1 x \bmod a_1 a_2 \leq dx$ and $ua_2 dx \bmod a_1 a_2 \leq dx$ have the same set of solutions. In order to see this, it is enough to see that $ua_2 d \equiv a_2 b_1 \bmod a_1 a_2$. But this is clear because $a_2 u \equiv 1 \bmod a_1$ implies that $a_2ub_1 \equiv b_1 \bmod a_1$, and so $(a_2b_1 - a_1b_2)u \equiv b_1 \bmod a_1$, that is, $du \equiv b_1 \bmod a_1$. Finally, by multiplying by a_2 we get $dua_2 \equiv a_2b_1 \bmod a_1a_2$. □

We can extract the following consequence from the proof of this last result.

Corollary 6.19. *Let a_1, a_2, b_1 and b_2 be positive integers such that $\frac{a_1}{b_1} < \frac{a_2}{b_2}$ and* $\gcd\{a_1,a_2\} = 1$. *Then*

$$\mathrm{S}\left(\left[\frac{a_1}{b_1}, \frac{a_2}{b_2}\right]\right) = \frac{\langle a_1, a_2\rangle}{a_2b_1 - a_1b_2}.$$

Open Problem 6.20. Let n_1 and n_2 be two positive relatively prime integers and let d be a positive integer.

1) Find a formula for the largest multiple of d that does not belong to $\langle n_1, n_2\rangle$.
2) Find a formula for the cardinality of the set of multiples of d that are not in $\langle n_1, n_2\rangle$.
3) Find a formula for the smallest multiple of d that belongs to $\langle n_1, n_2\rangle$.

The first and second problems are a generalization of the Frobenius problem solved by Sylvester in [102] (see Proposition 2.13); they propose to find a formula for the Frobenius number and genus of $\frac{\langle n_1, n_2\rangle}{d}$. The third problem asks about the multiplicity of $\frac{\langle n_1, n_2\rangle}{d}$.

As a consequence of Theorem 6.18 we obtain the following property.

Proposition 6.21. *Every numerical semigroup having a Toms decomposition can be expressed as a finite intersection of proportionally modular numerical semigroups.*

We will show that the converse also holds following the steps given in [84]. The key idea is contained in the following result, which is telling us that we can slightly modify the ends of the interval defining a proportionally modular numerical semigroup without modifying the semigroup.

Proposition 6.22. *Let α and β be rational numbers with $1 < \alpha < \beta$. Then there exist rational numbers $\overline{\alpha}$ and $\overline{\beta}$ such that $1 \leq \overline{\alpha} < \alpha < \beta < \overline{\beta}$ and $\mathrm{S}(]\overline{\alpha}, \overline{\beta}[) = \mathrm{S}([\alpha, \beta])$.*

Proof. Let $S = \mathrm{S}([\alpha, \beta])$. We know (by Lemma 5.8) that if $h \in \mathrm{G}(S)$, then there exists $n_h \in \mathbb{N}$ such that $\frac{h}{n_h+1} < \alpha < \beta < \frac{h}{n_h}$. The proof follows by choosing $\overline{\alpha} = \max\left\{\frac{h}{n_h+1} \mid h \in \mathrm{G}(S)\right\}$ and $\overline{\beta} = \min\left\{\frac{h}{n_h} \mid h \in \mathrm{G}(S)\right\}$. □

With this, we can prove that a proportionally modular numerical semigroup given by a closed interval can be represented by infinitely many quotients of embedding dimension two numerical semigroups.

Lemma 6.23. *Let a_1, a_2, b_1 and b_2 be positive integers such that $1 < \frac{a_1}{b_2} < \frac{a_2}{b_2}$. Then there exist positive integers a_0, b_0 and N such that $b_0 < a_0$ and for every integer $x \geq N$ with $\gcd\{x, a_2\} = 1$,*

$$\mathrm{S}\left(\left[\frac{a_1}{b_1},\frac{a_2}{b_2}\right]\right)=\frac{\langle x,a_2\rangle}{a_2\left\lfloor\frac{b_0x}{a_0}\right\rfloor-b_2x}.$$

Proof. By Proposition 6.22, we deduce that there exist positive integers a_0 and b_0 such that $1<\frac{a_0}{b_0}<\frac{a_1}{b_1}<\frac{a_2}{b_2}$, and if $\frac{a_0}{b_0}\leq\frac{x}{y}\leq\frac{a_1}{b_1}$ for some positive integers x and y, then $\mathrm{S}\left(\left[\frac{a_1}{b_1},\frac{a_2}{b_2}\right]\right)=\mathrm{S}\left(\left[\frac{x}{y},\frac{a_2}{b_2}\right]\right)$. Let N be the Frobenius number of $\mathrm{S}\left(\left[\frac{a_0}{b_0},\frac{a_1}{b_1}\right]\right)$ plus one. If x is an integer greater than or equal to N, then $x\in\mathrm{S}\left(\left[\frac{a_0}{b_0},\frac{a_1}{b_1}\right]\right)$, and by Lemma 5.8 there exists a positive integer y such that $\frac{a_0}{b_0}\leq\frac{x}{y}\leq\frac{a_1}{b_1}$. Hence $y\leq\frac{b_0x}{a_0}$. As y is an integer, $y\leq\left\lfloor\frac{b_0x}{a_0}\right\rfloor\leq\frac{b_0x}{a_0}$. We deduce that $\frac{a_0}{b_0}\leq x/\left\lfloor\frac{b_0x}{a_0}\right\rfloor\leq\frac{a_1}{b_1}$. Thus $\mathrm{S}\left(\left[\frac{a_1}{b_1},\frac{a_2}{b_2}\right]\right)=\mathrm{S}\left(\left[x/\left\lfloor\frac{b_0x}{a_0}\right\rfloor,\frac{a_2}{b_2}\right]\right)$. We now use Corollary 6.19 to conclude the proof.   □

Among the representations given in the preceding result, we can choose infinitely many that are Toms blocks as we see next.

Lemma 6.24. *Let a_1, a_2, b_1 and b_2 be positive integers such that $1<\frac{a_1}{b_1}<\frac{a_2}{b_2}$ and $\gcd(\{a_2,b_2\})=1$. Then there exist positive integers a_0, b_0 and N such that $b_0<a_0$ and for every integer $k\geq N$ one has that*

$$\mathrm{S}\left(\left[\frac{a_1}{b_1},\frac{a_2}{b_2}\right]\right)=\frac{\langle ka_0b_0a_2+1,a_2\rangle}{kb_0a_2(b_0a_2-b_2a_0)-b_2}.$$

Moreover, this is a Toms block.

Proof. Let $S=\mathrm{S}\left(\left[\frac{a_1}{b_1},\frac{a_2}{b_2}\right]\right)$. By Lemma 6.23, we know that there exist positive integers $b_0<a_0$ and N such that for all $x\geq N$ with $\gcd(\{x,a_2\})=1$, one has that $S=\frac{\langle x,a_2\rangle}{a_2\left\lfloor\frac{b_0x}{a_0}\right\rfloor-b_2x}$. Let $k\geq\frac{N-1}{a_0b_0a_2}$. Then $x=ka_0b_0a_2+1$ is greater than or equal to N, $\gcd(\{x,a_2\})=1$, and since $b_0<a_0$,

$$\left\lfloor\frac{b_0x}{a_0}\right\rfloor=\left\lfloor\frac{ka_0b_0^2a_2}{a_0}+\frac{b_0}{a_0}\right\rfloor=ka_2b_0^2.$$

Hence

$$S=\frac{\langle ka_0b_0a_2+1,a_2\rangle}{ka_2^2b_0^2-b_2(ka_0b_0a_2+1)}=\frac{\langle ka_0b_0a_2+1,a_2\rangle}{ka_2b_0(a_2b_0-a_0b_2)-b_2}.$$

Next we show that this representation is a Toms block.

- $\gcd(\{ka_0b_0a_2+1,a_2\})=1$,
- $\gcd(\{ka_2b_0(a_2b_0-a_0b_2)-b_2,a_2\})=\gcd(\{b_2,a_2\})=1$,
- $\gcd(\{ka_0b_0a_2+1,ka_2^2b_0^2-b_2(ka_0b_0a_2+1)\})=\gcd(\{ka_0b_0a_2+1,ka_2^2b_0^2\})=1$.
                                                                                      □

Now we see that in a family with finitely many proportionally modular numerical semigroups given by closed intervals, we can choose the denominator of the right ends to be the same for all the semigroups in the family.

Lemma 6.25. *For every* $i \in \{1,\ldots,r\}$, *let* $S_i = S\left(\left[\frac{a_{i,1}}{b_{i,1}}, \frac{a_{i,2}}{b_{i,2}}\right]\right)$ *with* $a_{i,1}$, $a_{i,2}$, $b_{i,1}$ *and* $b_{i,2}$ *positive integers with* $1 < \frac{a_{i,1}}{b_{i,1}} < \frac{a_{i,2}}{b_{i,2}}$. *Then there exist positive integers* $c_1,\ldots,c_r$ *and* d *such that for all* $i \in \{1,\ldots,r\}$,

- $S_i = S\left(\left[\frac{a_{i,1}}{b_{i,1}}, \frac{c_i}{d}\right]\right)$ *and*
- $\gcd\{c_i, d\} = 1$.

Proof. The sequence $\{\frac{ka_{i,2}+1}{kb_{i,2}}\}_{k \in \mathbb{N}\setminus\{0\}}$ is strictly decreasing and converges to $\frac{a_{i,2}}{b_{i,2}}$. Thus, in view of the proof of Proposition 6.22, there exists $N_i \in \mathbb{N}$ such that if $k_i \geq N_i$, we have that $S_i = S\left(\left[\frac{a_{i,1}}{b_{i,1}}, \frac{k_i a_{i,2}+1}{k_i b_{i,2}}\right]\right)$. To conclude the proof it suffices to choose $k_i = t\frac{b_{1,2}^2 \cdots b_{r,2}^2}{b_{i,2}}$ with t large enough so that $k_i \geq N_i$ for all $i \in \{1,\ldots,r\}$. $\square$

Theorem 6.26. *A numerical semigroup has a Toms decomposition if and only if it is the intersection of finitely many proportionally modular numerical semigroups.*

Proof. We already know one implication (Proposition 6.21).

Let S be the intersection of finitely many proportionally modular numerical semi-groups. If $S = \mathbb{N}$, then $\mathbb{N} = \frac{\langle 2,3 \rangle}{5}$ suits our needs. Otherwise, $S = S_1 \cap \cdots \cap S_r$ for some $S_1,\ldots,S_r$ proportionally modular numerical semigroups different from $\mathbb{N}$. In view of Theorem 5.14 and Lemma 6.25, there exist some positive integers $a_1,\ldots,a_r$, $b_1,\ldots,b_r$, $c_1,\ldots,c_r$ and d such that $S_i = S\left(\left[\frac{a_i}{b_i}, \frac{c_i}{d}\right]\right)$ with $\gcd\{c_i, d\} = 1$ for all $i \in \{1,\ldots,r\}$. From Lemma 6.24, we know that there exist positive integers $b_{i_0} < a_{i_0}$ and $N_i \in \mathbb{N}$ such that for all $k_i \geq N_i$, one obtains that

$$\frac{\langle k_i a_{i_0} b_{i_0} c_i + 1, c_i \rangle}{k_i c_i b_{i_0}(c_i b_{i_0} - a_{i_0} d) - d}$$

is a Toms block equal to S_i. Let $m_i = c_i b_{i_0}(c_i b_{i_0} - a_{i_0} d)$. Let $t = \max\{N_1,\ldots,N_r\}$. Then setting $k_i = t\frac{m_1 \cdots m_r}{m_i}$ one concludes that

$$S = \bigcap_{i=1}^{r} \frac{\langle k_i a_{i_0} b_{i_0} c_i + 1, c_i \rangle}{t m_1 \cdots m_r - d}$$

is a Toms representation for S. $\square$

We end this section by giving some other consequences of Proposition 6.22.

Proposition 6.27. *Let* S *be a proportionally modular numerical semigroup other than* $\mathbb{N}$. *Then there exists a positive integer* N *such that if* x *and* y *are relatively prime integers greater than* N, *there exists a positive integer* z *such that* $S = \frac{\langle x,y \rangle}{z}$.

Proof. By Theorem 5.14, there exist positive rational numbers α and β with $1 < \alpha < \beta$ such that $S = S([\alpha, \beta])$. Let $\overline{\alpha}$ and $\overline{\beta}$ be the rational numbers whose existence is ensured by Proposition 6.22. Let $N = F(S(]\overline{\alpha}, \alpha]) \cap S([\beta, \overline{\beta}[)) + 1$. For every x and y greater than N, by Lemma 5.8, there exist positive integers u and v such that

$\overline{\alpha} < \frac{x}{u} \leq \alpha$ ($x \in S(]\overline{\alpha},\alpha])$) and $\beta \leq \frac{y}{v} < \overline{\beta}$ ($y \in S([\beta,\overline{\beta}[)$). As a consequence of the proof of Proposition 6.22, $S\left(\left[\frac{x}{u},\frac{y}{v}\right]\right) = S$. If we choose x and y relatively prime, then Corollary 6.19 ensures the existence of an integer z such that $S = \frac{\langle x,y \rangle}{z}$.   □

This result produces a surprising characterization of proportionally modular numerical semigroups more restrictive than Theorem 6.18.

Corollary 6.28 ([54]). *Every proportionally modular numerical semigroup is of the form $\frac{\langle a,a+1 \rangle}{d}$ for some positive integers a and d.*

Choosing a large enough, we get the following characterization of numerical semigroups having a Toms decomposition.

Corollary 6.29 ([54]). *A numerical semigroup has a Toms decomposition if and only if there exist positive integers $a, p_1, \ldots, p_r$ such that*

$$S = \frac{\langle a, a+1 \rangle}{p_1} \cap \cdots \cap \frac{\langle a, a+1 \rangle}{p_r}.$$

Exercises

Exercise 6.1 ([92]). A numerical semigroup is said to be *arithmetic* if there exist integers a and b with $0 \leq b < a$ such that $A = \langle a, a+1, \ldots, a+b \rangle$.

a) Prove that $x \in \langle a, a+1, \ldots, a+b \rangle$ if and only $x \bmod a \leq \lfloor \frac{x}{a} \rfloor b$ ([33]).
b) Show that if $S = S\left(\left[\frac{a}{c},\frac{b}{c}\right]\right)$ for some positive integers a, b and c, then $S = \langle a, a+1, \ldots, a+b \rangle / c$.
c) Prove that a numerical semigroup is proportionally modular if and only if it is the quotient of an arithmetic numerical semigroup by a positive integer.

Exercise 6.2 ([92]). Show that $\langle 4,6,7 \rangle$ has no Toms decomposition (*Hint:* Prove first that this semigroup is irreducible).

Exercise 6.3 ([98]). Let n_1, n_2 and t be positive integers such that $\gcd\{n_1,tn_2\} = 1$. Show that

$$\frac{\langle n_1,tn_2 \rangle}{td} = \frac{\langle n_1,n_2 \rangle}{d}.$$

Exercise 6.4 ([98]). Let n_1 and n_2 be odd relatively prime integers. Prove that

$$\frac{\langle n_1,n_2 \rangle}{2} = \left\langle n_1,n_2,\frac{n_1+n_2}{2} \right\rangle.$$

Exercise 6.5 ([98]). Let n_1 and n_2 be positive integers such that n_1, n_2 and 3 are relatively prime. Prove that

a) if $n_1 + n_2 \equiv 0 \bmod 3$, then $\frac{\langle n_1,n_2 \rangle}{3} = \langle n_1,n_2,\frac{n_1+n_2}{3} \rangle$,

b) if $n_1 + 2n_2 \equiv 0 \bmod 3$, then $\frac{\langle n_1, n_2 \rangle}{3} = \langle n_1, n_2, \frac{n_1 + 2n_2}{3}, \frac{2n_1 + n_2}{3} \rangle$.

Exercise 6.6 ([98]). Let n_1, n_2 and d be positive integers with $\gcd\{n_1, n_2\} = 1$. Prove that if d divides $n_1 n_2 - n_1 - n_2$, then $\frac{\langle n_1, n_2 \rangle}{d}$ is symmetric. Show with an example that the converse is not true.

Exercise 6.7 ([64]). Let n_1 and n_2 be coprime integers greater than or equal to three. Prove that

a) if n_2 is even, then $\mathrm{F}\left(\frac{\langle n_1, n_2 \rangle}{2}\right) = \frac{n_1 n_2 - 2n_1 - n_2}{2}$ and $\mathrm{g}\left(\frac{\langle n_1, n_2 \rangle}{2}\right) = \frac{(n_1 - 1)(n_2 - 2)}{4}$,

b) if n_1 and n_2 are odd, then $\mathrm{F}\left(\frac{\langle n_1, n_2 \rangle}{2}\right) = \frac{n_1 n_2 - n_1 - n_2 - \min\{n_1, n_2\}}{2}$ and $\mathrm{g}\left(\frac{\langle n_1, n_2 \rangle}{2}\right) = \frac{(n_1 - 1)(n_2 - 1)}{4}$.

Exercise 6.8 ([64]). Let n_1 and n_2 be coprime integers greater than or equal to three. Prove that

a) if n_2 is even, then $\#\mathrm{FG}(\langle n_1, n_2 \rangle) = \frac{n_2(n_1 - 1)}{4} - \left\lceil \frac{n_2}{6} \right\rceil \left\lceil \frac{n_1 - 3}{6} \right\rceil$,

b) if n_1 and n_2 are odd, then $\#\mathrm{FG}(\langle n_1, n_2 \rangle) = \frac{(n_1 - 1)(n_2 - 1)}{4} - \left\lceil \frac{(n_2 - 3)}{6} \right\rceil \left\lceil \frac{n_1 - 3}{6} \right\rceil$.

Exercise 6.9 ([64]). Let n_1 and n_2 be coprime integers greater than or equal to three. Let $S = \frac{\langle n_1, n_2 \rangle}{3}$. Denote by $\mathrm{F}_3(S)$ the largest multiple of 3 not belonging to S. Show that

a) if $n_2 \equiv 0 \bmod 3$, then $\mathrm{F}_3(S) = n_1 n_2 - 3n_1 - n_2$,

b) if n_1 and n_2 are congruent with 1 modulo 3, then $\mathrm{F}_3(S) = n_1 n_2 - n_1 - n_2 - \min\{n_1, n_2\}$,

c) if $n_1 \equiv 1 \bmod 3$ and $n_2 \equiv 2 \bmod 3$, then $\mathrm{F}_3(S) = n_1 n_2 - n_1 - n_2 - \min\{2n_1, n_2\}$,

d) if n_1 and n_2 are congruent with 2 modulo 3, then $\mathrm{F}(S) = n_1 n_2 - n_1 - n_2$.

Exercise 6.10 ([24]). Let $S = \langle 4, 6, 7, 9 \rangle$. Prove that

a) S admits a Toms decomposition,

b) $S \setminus \{6\}$ also admits a Toms decomposition,

c) $S \setminus \{9\}$ does not admit a Toms decomposition.

Exercise 6.11. Let S be a proportionally modular numerical semigroup minimally generated by $\{n_1, \ldots, n_p\}$, where these generators are arranged in a way that there exist positive integers $d_1, \ldots, d_p$ such that $\frac{n_1}{d_1} < \cdots < \frac{n_p}{d_p}$ is a proper Bézout sequence with adjacent edges (Proposition 5.33). Assume that $h \in \{1, \ldots, p\}$ is such that $n_1 > \cdots > n_h < \cdots < n_p$ (Proposition 5.29).

a) Prove that for all $i \in \{1, \ldots, p\}$, $\frac{n_i}{d_i + 1} < \frac{n_1}{d_1}$ and if $d_i \neq 0$, then $\frac{n_i}{d_i - 1} > \frac{n_p}{d_p}$.

b) Let $i \in \{1, \ldots, h - 1\}$. Prove that $d_i \neq 1$ and that $S \subseteq \mathrm{S}\left(\left] \frac{n_i}{d_i}, \frac{n_i}{d_i - 1} \right[\right)$.

c) Let $i \in \{h, \ldots, p\}$. Prove that $S \subseteq \mathrm{S}\left(\left] \frac{n_i}{d_i + 1}, \frac{n_i}{d_i} \right[\right)$.

d) Show that for all $i \in \{1, \ldots, p\}$, $S \setminus \{n_i\}$ admits a Toms decomposition.

e) Prove that for all $\{i_1, \ldots, i_k\} \subseteq \{1, \ldots, p\}$, S admits a Toms decomposition.

Exercise 6.12 ([54]). Let α and β be rational numbers with $0 < \alpha < \beta$. Prove that there exist a positive rational number ε such that for all rational numbers δ with $0 \le \delta < \varepsilon$, the equality $S([\alpha - \delta, \beta + \delta]) = S([\alpha, \beta])$ holds.

Exercise 6.13 ([54]). We say that two systems of proportionally modular Diophantine inequalities are equivalent if they have the same set of integer solutions. Prove that any system of proportionally modular Diophantine inequalities is equivalent to a system of proportionally modular Diophantine inequalities in which any inequality has the same modulus and furthermore this modulus can be chosen to be prime.

Chapter 6
Families of numerical semigroups closed under finite intersections and adjoin of the Frobenius number

Introduction

The concept of Frobenius variety was introduced in [67] in order to unify most of the results appearing in [88, 89, 24, 11].

In this chapter we show that some families appearing previously in this book are Frobenius varieties. We will introduce new varieties, as for instance, those fulfilling a pattern and those families generated by an arbitrary set of numerical semigroups.

We also show how the numerical semigroups belonging to a Frobenius variety can be arranged in a directed acyclic graph, and will paint part of this graph for some known Frobenius varieties.

While working with numerical semigroups in a Frobenius variety we can represent these semigroups in the classic way by their minimal systems of generators. By doing this we are not taking advantage of the fact that these semigroups belong to a certain Frobenius variety. In order to avoid this, for a Frobenius variety V, we introduce the concept of minimal V-system of generators.

1 The directed graph of the set of numerical semigroups

We see how the set of numerical semigroups can be arranged in a tree. This will enable us to construct recursively the set of all numerical semigroups. We use this idea later in this chapter for certain families of numerical semigroups.

A *directed graph* G is a pair (V, E), where V is a nonempty set whose elements are called *vertices*, and E is a subset of $\{(v, w) \in V \times V \mid v \neq w\}$. The elements of E are called *edges* of G. A *path* connecting the vertices x and y of G is a sequence of distinct edges of the form $(v_0, v_1), (v_1, v_2), \ldots, (v_{n-1}, v_n)$ with $v_0 = x$ and $v_n = y$. A graph G is a *tree* if there exists a vertex r (known as the *root* of G) such that for every other vertex x of G, there exists a unique path connecting x and r. If (x, y) is

J.C. Rosales, P.A. García-Sánchez, *Numerical Semigroups,*
Developments in Mathematics 20, DOI 10.1007/978-1-4419-0160-6_7,
© Springer Science+Business Media, LLC 2009

an edge of a tree, then we say that x is a *son* of y. A vertex in a tree is a *leaf* if it has no sons.

Let $\mathscr{S}$ be the set of numerical semigroups. We define the directed graph $\mathscr{G}(\mathscr{S})$ as the graph whose vertices are the elements of $\mathscr{S}$ and $(T,S) \in \mathscr{S} \times \mathscr{S}$ is an edge if $S = T \cup \{F(T)\}$.

Clearly, the candidate of root for $\mathscr{G}(\mathscr{S})$ is $\mathbb{N}$. The path connecting a numerical semigroup S with $\mathbb{N}$ can be defined by means of the following sequence:

- $S_0 = S$,
- $S_{i+1} = \begin{cases} S_i \cup \{F(S_i)\} & \text{if } S_i \neq \mathbb{N}, \\ \mathbb{N} & \text{otherwise.} \end{cases}$

As the complement of S in $\mathbb{N}$ is finite, there exists a positive integer k such that $S_k = \mathbb{N}$. Clearly

$$S = S_0 \subsetneq S_1 \subsetneq \cdots \subsetneq S_k = \mathbb{N}$$

provides a path connecting S and $\mathbb{N}$ in the graph $\mathscr{G}(\mathscr{S})$. We define

$$\mathrm{Ch}(S) = \{S_0, S_1, \ldots, S_k\}$$

and will refer to this set as the *chain* associated to S.

Note also that if $S = T \cup \{F(T)\}$, then $F(T)$ becomes a minimal generator of S, which in addition is greater than the Frobenius number of S. Conversely, if we choose a minimal generator n of S, then $S \setminus \{n\}$ is a numerical semigroup, and if this generator is greater than the Frobenius number, then $F(S \setminus \{n\}) = n$. With all this information the following property is easy to prove.

Proposition 7.1. *The directed graph $\mathscr{G}(\mathscr{S})$ is a tree rooted in $\mathbb{N}$. Moreover, the sons of $S \in \mathscr{S}$ are $S \setminus \{x_1\}, \ldots, S \setminus \{x_r\}$, with $x_1, \ldots, x_r$ those minimal generators greater than $F(S)$.*

This result can be used to recurrently construct the tree, starting in $\mathbb{N}$, containing the set of all numerical semigroups. Clearly, a leaf is a semigroup all of whose minimal generators are smaller than the Frobenius number.

Proposition 7.2. *Let $S \in \mathscr{S}$. Then S is a leaf of $\mathscr{G}(\mathscr{S})$ if and only if it does not have minimal generators greater than $F(S)$.*

Example 7.3. We draw three levels of $\mathscr{G}(\mathscr{S})$.

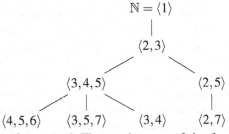

The semigroup $\langle 3,4 \rangle$ is a leaf. The semigroups of the form $\langle 2,k \rangle$ with $k > 3$ always have exactly one son.

It is also easy to prove from the definitions the following proposition.

Proposition 7.4. *If (T,S) is an edge of $\mathscr{G}(\mathscr{S})$, then*

- $F(T) > F(S)$,
- $g(T) = g(S) + 1$.

With this result, we can use the recurrent construction of $\mathscr{G}(\mathscr{S})$ to determine all numerical semigroups with Frobenius number (or genus) less than a given amount.

2 Frobenius varieties

Recall that the intersection of two numerical semigroups is again a numerical semigroup (Exercise 2.2). If S is a numerical semigroup, then by Lemma 4.1, $S \cup \{F(S)\}$ is also a numerical semigroup. In the rest of this chapter we will make use of these two results without quoting them.

A *Frobenius variety* is a nonempty set V of numerical semigroups fulfilling the following conditions

1) if S and T are in V, then so is $S \cap T$,
2) if S is in V and $S \neq \mathbb{N}$, then $S \cup \{F(S)\} \in V$.

Clearly the set $\mathscr{S}$ of all numerical semigroups is a Frobenius variety. The chain associated to a numerical semigroup is also a Frobenius variety. We review in this section some other notable examples of Frobenius varieties.

Let A be a subset of $\mathbb{N}$. The set $\{S \in \mathscr{S} \mid A \subseteq S\}$ is a Frobenius variety. In particular, $\mathscr{O}(S)$, the set of oversemigroups of S, is a Frobenius variety.

2.1 Arf numerical semigroups

Recall that a numerical semigroup S has the Arf property if for any $x, y, z \in S$ with $x \geq y \geq z$, $x + y - z \in S$. We already know (Proposition 3.22) that the set of Arf numerical semigroups is closed under finite intersections. In order to prove that this family is a Frobenius variety, we must prove the following result.

Lemma 7.5. *Let S be an Arf numerical semigroup other than $\mathbb{N}$. Then $S \cup \{F(S)\}$ is also an Arf numerical semigroup.*

Proof. Take $x, y, z \in S \cup \{F(S)\}$ such that $x \geq y \geq z$, and let us prove that $x + y - z \in S \cup \{F(S)\}$.

- If $x, y, z \in S$, then as S is Arf, we obtain that $x + y - z \in S \subset S \cup \{F(S)\}$.
- If $F(S) \in \{x, y, z\}$, then $x + y - z \geq F(S)$ and thus $x + y - z \in S \cup \{F(S)\}$. □

As we have mentioned above, this together with Proposition 3.22 proves that the set of Arf numerical semigroups is a Frobenius variety.

Proposition 7.6. *The set of Arf numerical semigroups is a Frobenius variety.*

2.2 Saturated numerical semigroups

Now we see that the same holds for the set of saturated numerical semigroups. We proceed analogously. Proposition 3.39 asserts that the intersection of finitely many saturated numerical semigroups is again saturated (this fact follows easily from the definition, as was already mentioned before stating Proposition 3.39). Thus once more, proving that the family of saturated numerical semigroups is a Frobenius variety passes through demonstrating the following lemma.

Lemma 7.7. *Let $S \neq \mathbb{N}$ be a saturated numerical semigroup. Then $S \cup \{F(S)\}$ is also saturated.*

Proof. Let $T = S \cup \{F(S)\}$. In view of Proposition 3.34, it suffices to show that if $s \in T$, then $s + d_T(s) \in T$. If $s < F(S)$, then $s \in S$ and $d_S(s) = d_T(s)$, whence $s + d_T(s) = s + d_S(s) \in S \subset T$. If $s \geq F(S)$, then $s + d_T(s) \geq F(S)$, and thus $s + d_T(s) \in T$. $\square$

Proposition 7.8. *The set of saturated numerical semigroups is a Frobenius variety.*

2.3 Numerical semigroups having a Toms decomposition

The set of numerical semigroups having a Toms decomposition coincides with the set of numerical semigroups that are intersections of finitely many proportionally modular numerical semigroups (Theorem 6.26). Clearly, this family is closed under finite intersection. So it suffices to show that this family is also closed under the adjoin of the Frobenius number to see that it is a Frobenius variety. Following [24], we start proving that adding the Frobenius number to a proportionally modular numerical semigroup yields a numerical semigroup that is the intersection of finitely many proportionally modular numerical semigroups.

Lemma 7.9. *Let S be a proportionally modular numerical semigroup with $S \neq \mathbb{N}$. Then $S \cup \{F(S)\}$ is the intersection of finitely many proportionally modular numerical semigroups.*

Proof. Assume that S is generated by $\{n_1, \ldots, n_p\}$. Since S is proportionally modular, $S = S([\alpha, \beta])$ for some real numbers α, β with $1 < \alpha < \beta$ (Theorem 5.14). By Lemma 5.8, for every $i \in \{1, \ldots, p\}$, there exists $d_i \in \{1, \ldots, n_i - 1\}$ such that $\frac{n_i}{d_i} \in [\alpha, \beta]$. Assume that after rearranging the generators (if needed), we have that

$$\frac{n_1}{d_1} < \cdots < \frac{n_p}{d_p}.$$

Then

$$S\left(\left[\frac{n_1}{d_1}, \frac{n_p}{d_p}\right]\right) \subseteq S([\alpha, \beta]) = S,$$

and as $n_1, \ldots, n_p \in S([\frac{n_1}{d_1}, \frac{n_p}{d_p}])$ (Lemma 5.8 again), we deduce that

$$S = S\left(\left[\frac{n_1}{d_1}, \frac{n_p}{d_p}\right]\right).$$

Since $F(S) \notin S$, by using Lemma 5.8 once more, there exists $d \in \mathbb{N}$ such that

$$\frac{F(S)}{d+1} < \frac{n_1}{d_1} < \cdots < \frac{n_p}{d_p} < \frac{F(S)}{d}.$$

If $d = 0$, then $F(S) < \frac{n_i}{d_i}$, and thus $F(S) < n_i$ for all $i \in \{1, \ldots, p\}$. Hence $F(S) < s$ for all $s \in S \setminus \{0\}$, which means that $S = \{0, F(S)+1, \rightarrow\}$. In this setting $S \cup \{F(S)\} = \{0, F(S), \rightarrow\} = S([F(S), 2F(S)])$, which is proportionally modular by Theorem 5.14.

Now assume that $d \neq 0$. We prove that

$$S \cup \{F(S)\} = S\left(\left[\frac{F(S)}{d+1}, \frac{n_p}{d_p}\right]\right) \cap S\left(\left[\frac{n_1}{d_1}, \frac{F(S)}{d}\right]\right),$$

which in view of Theorem 5.14 proves that S is the intersection of proportionally modular numerical semigroups. The inclusion

$$S \cup \{F(S)\} \subseteq S\left(\left[\frac{F(S)}{d+1}, \frac{n_p}{d_p}\right]\right) \cap S\left(\left[\frac{n_1}{d_1}, \frac{F(S)}{d}\right]\right)$$

is clear. Now, assume that there exists $x \in S([\frac{F(S)}{d+1}, \frac{n_p}{d_p}]) \cap S([\frac{n_1}{d_1}, \frac{F(S)}{d}])$, with $x \notin S \cup \{F(S)\}$. Thus $x < F(S)$ and there exist n and m positive integers such that

$$\frac{F(S)}{d+1} \leq \frac{x}{n} \leq \frac{n_p}{d_p} \quad \text{and} \quad \frac{n_1}{d_1} \leq \frac{x}{m} \leq \frac{F(S)}{d}$$

(Lemma 5.8). As $\frac{x}{n}, \frac{x}{m} \notin [\frac{n_1}{d_1}, \frac{n_p}{d_p}]$ (this would imply that $x \in S$), we have that

$$\frac{F(S)}{d+1} \leq \frac{x}{n} < \frac{n_1}{d_1} < \frac{n_p}{d_p} < \frac{x}{m} \leq \frac{F(S)}{d},$$

which in particular implies that $m < n$. From the preceding inequalities we deduce that

$$F(S)n \leq xd + x \leq F(S)m + x < F(S)m + F(S) = F(S)(m+1),$$

and this leads to $n < m+1$. We conclude that $m < n < m+1$, which is impossible.

$\square$

With this result it is now easy to show the following.

Lemma 7.10. *Let S be a numerical semigroup. If $S \neq \mathbb{N}$ and S is the intersection of finitely many proportionally modular numerical semigroups, then $S \cup \{F(S)\}$ is also the intersection of finitely many proportionally modular numerical semigroups.*

Proof. Assume that there exist $S_1, \ldots, S_t$ proportionally modular numerical semigroups such that $S = S_1 \cap \cdots \cap S_t$. Then $S \cup \{F(S)\} = (S_1 \cup \{F(S)\}) \cap \cdots \cap (S_t \cup \{F(S)\})$. For $i \in \{1, \ldots, t\}$, if $F(S) \in S_i$, then $S_i \cup \{F(S)\} = S_i$. If $F(S) \notin S_i$, as $S \subseteq S_i$, this implies that $F(S) = F(S_i)$. In view of Lemma 7.9, $S_i \cup \{F(S_i)\}$ is the intersection of finitely many proportionally modular numerical semigroups. Thus $S \cup \{F(S)\}$ is also the intersection of finitely many proportionally modular numerical semigroups. $\square$

With the use of Theorem 6.26 the contents of this section read as follows.

Proposition 7.11. *The set of numerical semigroups having a Toms decomposition is a Frobenius variety.*

2.4 Numerical semigroups defined by patterns

The idea of pattern on a numerical semigroup was introduced in [11] in order to generalize the concept of Arf numerical semigroups. We see in this section that the set of numerical semigroups defined by certain patterns is a Frobenius variety.

A *pattern* P of length n is an expression of the form $a_1 x_1 + \cdots + a_n x_n$ with $x_1, \ldots, x_n$ unknowns and $a_1, \ldots, a_n$ nonzero integers. We say that a semigroup S *admits* the pattern P if for every $s_1, \ldots, s_n \in S$ with $s_1 \geq s_2 \geq \cdots \geq s_n$, the element $a_1 s_1 + \cdots + a_n s_n$ belongs to S. We denote by $\mathscr{S}(P)$ the set of all numerical semigroups admitting the pattern P.

Remark 7.12. The set $\mathscr{S}(x_1 + x_2 - x_3)$ is the set of all Arf numerical semigroups.

From the definition it is easy to see that if P is a pattern and $S_1, \ldots, S_r$ are numerical semigroups admitting this pattern, then the intersection $S_1 \cap \cdots \cap S_r$ also admits the pattern P.

Lemma 7.13. *Let P be a pattern. The set $\mathscr{S}(P)$ is closed under finite intersections.*

We next see what are the patterns for which $\mathscr{S}(P)$ is also closed under the adjunction of $F(S)$, or in view of the preceding lemma, for which $\mathscr{S}(P)$ is a Frobenius variety. To this end, we use the results given in [11]. First we see for which patterns the set $\mathscr{S}(P)$ is not empty.

Lemma 7.14. *Let $P = a_1 x_1 + \cdots + a_n x_n$ be a pattern. The following conditions are equivalent.*

1) $\mathbb{N} \in \mathscr{S}(P)$.

2) $\mathscr{S}(P)$ is not empty.

3) $\sum_{i=1}^{n'} a_i \geq 0$ for all $n' \leq n$.

Proof. Clearly *1)* implies *2)*.

Let us prove that *2)* implies *3)*. Assume to the contrary that there exists $n' \leq n$ such that $\sum_{i=1}^{n'} a_i < 0$. Let S be a numerical semigroup in $\mathscr{S}(P)$, and let l be a nonzero element of S. Take $s_1 = s_2 = \cdots = s_{n'} = l$ and $s_{n'+1} = \cdots = s_n = 0$. It is obvious that $\sum_{i=1}^{n} a_i s_i < 0$ and thus it is not in S, a contradiction.

Finally we see that *3)* implies *1)*. We must show that if $s_1, \ldots, s_n$ are nonnegative integers such that $s_1 \geq s_2 \geq \cdots \geq s_n$, then $a_1 s_1 + \cdots + a_n s_n \in \mathbb{N}$. But this is clear because $a_1 s_1 + \cdots + a_n s_n \geq a_1 s_n + a_2 s_n + \cdots + a_n s_n = (\sum_{i=1}^{n} a_i) s_n$ which is nonnegative by hypothesis. $\quad\square$

We say that a pattern P is *admissible* if $\mathscr{S}(P)$ is not empty. From the above characterization it easily follows that $a_1 \geq 0$ if $P = a_1 x_1 + \cdots + a_n x_n$ is an admissible pattern.

Given a pattern $P = a_1 x_1 + \cdots + a_n x_n$ define

$$P' = \begin{cases} (a_1 - 1)x_1 + a_2 x_2 + \cdots + a_n x_n & \text{if } a_1 > 1, \\ a_2 x_2 + \cdots + a_n x_n & \text{otherwise.} \end{cases}$$

The pattern P is *strongly admissible* if both P and P' are admissible patterns. As we see next for these patterns, $\mathscr{S}(P)$ is a Frobenius variety. In order to show that $S \in \mathscr{S}(P)$ implies $S \cup \{F(S)\} \in \mathscr{S}(P)$, we need a technical lemma.

Lemma 7.15. *Let $P = a_1 x_1 + \cdots + a_n x_n$ be a strongly admissible pattern of length n. Then for every $k_1 \geq \cdots \geq k_n$, it holds that $a_1 k_1 + \cdots + a_n k_n \geq k_1$.*

Proof. Since P' is admissible, we have that

$$P'(k_1, \ldots, k_n) \geq 0$$

$(P'(k_2, \ldots, k_n) \geq 0$, if $a_1 = 1)$, which leads to

$$P(k_1, \ldots, k_n) = k_1 + P'(k_1, \ldots, k_n) \geq k_1$$

$(P(k_1, \ldots, k_n) = k_1 + P'(k_2, \ldots, k_n) \geq k_1$, if $a_1 = 1)$. $\quad\square$

Now it is easy to show that $\mathscr{S}(P)$ is closed under the adjoin of the Frobenius number for P a strongly admissible pattern.

Lemma 7.16. *Let S be in $\mathscr{S}(P) \setminus \{\mathbb{N}\}$ with P a strongly admissible pattern. Then $S \cup \{F(S)\} \in \mathscr{S}(P)$.*

Proof. Assume that P has length n and let $s_1, \ldots, s_n$ be elements in $S \cup \{F(S)\}$ such that $s_1 \geq \cdots \geq s_n$. We wonder if $P(s_1, \ldots, s_n) \in S \cup \{F(S)\}$. We distinguish two cases.

- If $F(S) > s_1$, then $\{s_1, \ldots, s_n\} \subseteq S$. As $S \in \mathscr{S}(P)$, it follows that

$$P(s_1, \ldots, s_n) \in S \subset S \cup \{F(S)\}.$$

- If $F(S) \leq s_1$, then by Lemma 7.15, $P(s_1, \ldots, s_n) \geq s_1 \geq F(S)$ and thus

$$P(s_1, \ldots, s_n) \in S \cup \{F(S)\}. \qquad \qquad \square$$

Gathering all this information we obtain the result mentioned above.

Proposition 7.17. *Let P be a strongly admissible pattern. Then $\mathscr{S}(P)$ is a Frobenius variety.*

Example 7.18. Note that $x_1 + x_2 + x_3 - x_4$ is a strongly admissible pattern. Clearly $\mathscr{S}(x_1 + x_2 + x_3 - x_4)$ contains the set of all Arf numerical semigroups. However, $\langle 3, 4 \rangle \in \mathscr{S}(x_1 + x_2 + x_3 - x_4)$ and $\langle 3, 4 \rangle$ is not Arf (it does not have maximal embedding dimension).

3 Intersecting Frobenius varieties

Due to the following property we can obtain new Frobenius varieties by intersecting those appearing in the preceding sections.

Proposition 7.19. *The intersection of Frobenius varieties is a Frobenius variety.*

Proof. Let $\{V_i\}_{i \in I}$ be a family of Frobenius varieties. Observe that $\mathbb{N}$ belongs to any Frobenius variety. Hence $\bigcap_{i \in I} V_i$ is a nonempty family of numerical semigroups. If $S, S' \in \bigcap_{i \in I} V_i$, then $S, S' \in V_i$ for all $i \in I$. So we have that $S \cap S' \in V_i$ for all $i \in I$ and thus $S \cap S' \in \bigcap_{i \in I} V_i$. If $S \in \bigcap_{i \in I} V_i$ and $S \neq \mathbb{N}$, then $S \cup \{F(S)\} \in V_i$ for all $i \in I$ and therefore $S \cup \{F(S)\} \in \bigcap_{i \in I} V_i$. $\qquad \qquad \square$

This result is not only useful to construct new Frobenius varieties from known varieties, it also allows us to talk about the Frobenius variety *generated* by a family X of numerical semigroups. We denote this variety by $\mathscr{F}(X)$ and define it as the intersection of all Frobenius varieties containing X.

Given a family X of numerical semigroups, we denote by

$$\mathrm{Ch}(X) = \bigcup_{S \in X} \mathrm{Ch}(S).$$

Theorem 7.20. *Let X be a nonempty family of numerical semigroups. Then $\mathscr{F}(X)$ is the set of all finite intersections of elements in $\mathrm{Ch}(X)$.*

Proof. Let

$$V = \{ S_1 \cap \cdots \cap S_n \mid n \in \mathbb{N} \setminus \{0\} \text{ and } S_1, \ldots, S_n \in \mathrm{Ch}(X) \}.$$

By using that $\mathscr{F}(X)$ is a Frobenius variety containing X, we deduce that $V \subseteq \mathscr{F}(X)$. We see now that $\mathscr{F}(X) \subseteq V$. It suffices to prove that V is a Frobenius variety containing X and then apply that $\mathscr{F}(X)$ is the smallest Frobenius variety containing X. Clearly, if $S, S' \in V$, then $S \cap S' \in V$. Now we prove that if $S \in V$ and

$S \neq \mathbb{N}$, then $S \cup \{F(S)\} \in V$. Since $S \in V$, there exist $S_1, \ldots, S_n \in \mathrm{Ch}(X)$ such that $S = S_1 \cap \cdots \cap S_n$. It is easy to prove that $F(S) = \max\{F(S_1), \ldots, F(S_n)\}$. Hence $F(S_i) \leq F(S)$ for all $i \in \{1, \ldots, n\}$. For $i \in \{1, \ldots, n\}$, if $F(S) > F(S_i)$, we have $S_i \cup \{F(S)\} = S_i$, and if $F(S) = F(S_i)$, we have $S_i \cup \{F(S)\} = S_i \cup \{F(S_i)\}$. Hence, we obtain that for every $i \in \{1, \ldots, n\}$, $S_i \cup \{F(S)\} \in \mathrm{Ch}(X)$. From the equality $S \cup \{F(S)\} = (S_1 \cup \{F(S)\}) \cap \cdots \cap (S_n \cup \{F(S)\})$, we can assert that $S \cup \{F(S)\} \in V$. $\qquad\square$

4 Systems of generators with respect to a Frobenius variety

In all the examples of Frobenius varieties given so far, the concept of closure of a numerical semigroup (or of a subset of nonnegative integers) can be defined as the smallest (with respect to set inclusion) semigroup in the variety containing the given semigroup (or subset of nonnegative integers). Since these varieties are closed under intersections, this is the intersection of all the elements of the variety containing the given set. From this idea one can define the concept of system of generators with respect to the variety. These were studied for Arf semigroups in [88], for saturated semigroups in [89], for semigroups that are the intersection of proportionally modular numerical semigroups in [24], and for numerical semigroups defined by strongly admissible patterns in [11]. This idea was then abstracted and generalized to any Frobenius variety in [67].

In this section, $\mathscr{V}$ denotes a Frobenius variety. A submonoid M of $\mathbb{N}$ is a $\mathscr{V}$-*monoid* if it can be expressed as the intersection of elements of $\mathscr{V}$. From the definition it easily follows that the intersection of $\mathscr{V}$-monoids is a $\mathscr{V}$-monoid.

Lemma 7.21. *The intersection of $\mathscr{V}$-monoids is a $\mathscr{V}$-monoid.*

In view of this result, given a subset of nonnegative integers A, we can define the $\mathscr{V}$-monoid *generated* by A as the intersection of all $\mathscr{V}$-monoids containing A. We will denote this $\mathscr{V}$-monoid by $\mathscr{V}(A)$, and we say that A is a $\mathscr{V}$-*system of generators* of $\mathscr{V}(A)$. If no proper subset of A generates this $\mathscr{V}$-monoid, then we say that A is a *minimal $\mathscr{V}$-system* of generators.

Our aim in this section is to prove that minimal $\mathscr{V}$ systems of generators are unique and have finitely many elements.

The following properties are a direct consequence of the definitions.

Lemma 7.22.
1) Let A and B be subsets of $\mathbb{N}$. If $A \subseteq B$, then $\mathscr{V}(A) \subseteq \mathscr{V}(B)$.
2) For any subset A of nonnegative integers, $\mathscr{V}(A) = \mathscr{V}(\langle A \rangle)$.
3) If M is a $\mathscr{V}$-monoid, then $\mathscr{V}(M) = M$.

In view of these properties, we can deduce that every $\mathscr{V}$-monoid admits a finite system of generators, because we can always make use of the classic minimal system of generators of the underlying monoid (which is finite by Corollary 2.8).

Proposition 7.23. *Every $\mathscr{V}$-monoid has a $\mathscr{V}$-system of generators with finitely many elements.*

Minimal $\mathscr{V}$-systems of generators can be characterized as in the classic sense.

Lemma 7.24. *Let $A \subseteq \mathbb{N}$ and $M = \mathscr{V}(A)$. The set A is a minimal $\mathscr{V}$-system of generators of M if and only if $a \notin \mathscr{V}(A \setminus \{a\})$ for all $a \in A$.*

Proof. If $a \in \mathscr{V}(A \setminus \{a\})$, then $A \subseteq \mathscr{V}(A \setminus \{a\})$ and by Lemma 7.22 we obtain that $M = \mathscr{V}(A) \subseteq \mathscr{V}(\mathscr{V}(A \setminus \{a\})) = \mathscr{V}(A \setminus \{a\}) \subseteq \mathscr{V}(A) = M$. Thus, $M = \mathscr{V}(A \setminus \{a\})$, which implies that A is not a minimal $\mathscr{V}$-system of generators of M.

Conversely, if A is not a minimal $\mathscr{V}$-system of generators of M, then there exists a proper subset B of A such that $\mathscr{V}(B) = M$. Let $a \in A \setminus B$; applying again Lemma 7.22 we have that $a \in M = \mathscr{V}(B) \subseteq \mathscr{V}(A \setminus \{a\})$. Hence $a \in \mathscr{V}(A \setminus \{a\})$. $\square$

The proof of the fact that minimal $\mathscr{V}$-systems of generators are unique relies on the following result. If M is a submonoid of $\mathbb{N}$ and $x \in M$, then x can be expressed as a linear combination of those generators of M smaller than or equal to x. This observation that is obvious for submonoids of $\mathbb{N}$ needs to be clarified for $\mathscr{V}$-monoids.

Lemma 7.25. *Let $A \subseteq \mathbb{N}$. If $x \in \mathscr{V}(A)$, then $x \in \mathscr{V}(\{a \in A \mid a \leq x\})$.*

Proof. If $x \notin \mathscr{V}(\{a \in A \mid a \leq x\})$, then from the definition we deduce that there exists $S \in \mathscr{V}$ such that $\{a \in A \mid a \leq x\} \subseteq S$ and $x \notin S$. As $\mathscr{V}$ is a Frobenius variety and $S \in \mathscr{V}$ the semigroup $S \cup \{x+1, \rightarrow\}$ is also in $\mathscr{V}$ (this semigroup belongs to $\mathrm{Ch}(S)$). Clearly, $A \subseteq S \cup \{x+1, \rightarrow\}$ and $x \notin S \cup \{x+1, \rightarrow\}$. Hence $x \notin \mathscr{V}(A) = \bigcap_{S \in \mathscr{V}, A \subseteq S} S$.
                                                                                              $\square$

Theorem 7.26. *Let $\mathscr{V}$ be a Frobenius variety. Let M be a $\mathscr{V}$-monoid, and let A and B be two minimal $\mathscr{V}$-systems of generators of M. Then $A = B$.*

Proof. Assume that $A = \{a_1 < a_2 < \cdots\}$ and $B = \{b_1 < b_2 < \cdots\}$. If $A \neq B$, then there exists $i = \operatorname{minimum}\{k \mid a_k \neq b_k\}$. Without loss of generality we can assume that $a_i < b_i$. By using that $a_i \in M = \mathscr{V}(A) = \mathscr{V}(B)$, from Lemma 7.25 we deduce that $a_i \in \mathscr{V}(\{b_1, \ldots, b_{i-1}\})$. Now by the minimality of i we know that $\{b_1, \ldots, b_{i-1}\} = \{a_1, \ldots, a_{i-1}\}$. Thus $a_i \in \mathscr{V}(\{a_1, \ldots, a_{i-1}\})$. From Lemma 7.22 we have that $a_i \in \mathscr{V}(A \setminus \{a_i\})$, which is impossible in view of Lemma 7.24. $\square$

Summarizing we obtain that minimal $\mathscr{V}$-systems of generators are finite and unique as are classic minimal systems of generators of submonoids of $\mathbb{N}$.

Corollary 7.27. *Every $\mathscr{V}$-monoid has a unique minimal $\mathscr{V}$-system of generators and this set is finite.*

5 The directed graph of a Frobenius variety

Let $\mathscr{V}$ be a Frobenius variety. We present here a similar construction to the one already presented in Section 1 to build the graph of all numerical semigroups in $\mathscr{V}$.

First we focus on the calculation of the sons of a given element in $\mathscr{V}$. The result obtained is analogous to the classic one, but substituting minimal generators by minimal $\mathscr{V}$-generators.

Proposition 7.28. *Let M be a $\mathscr{V}$-monoid and let $x \in M$. The set $M \setminus \{x\}$ is a $\mathscr{V}$-monoid if and only if x belongs to the minimal $\mathscr{V}$-system of generators of M.*

Proof. Let A be the minimal $\mathscr{V}$-system of generators of M and let $x \in M$. We see that if $M \setminus \{x\}$ is a $\mathscr{V}$-monoid, then $x \in A$. If $x \notin A$, then $A \subseteq M \setminus \{x\}$. Since $M \setminus \{x\}$ is a $\mathscr{V}$-monoid containing A, we have that $M = \mathscr{V}(A) \subseteq M \setminus \{x\}$, a contradiction.

Conversely, if x belongs to A, then by Theorem 7.26 we deduce that $\mathscr{V}(M \setminus \{x\}) \neq \mathscr{V}(A) = M$, since otherwise from $M \setminus \{x\}$ we could find a minimal system of generators contained in $M \setminus \{x\}$ and thus not equal to A. Hence $M \setminus \{x\} \subseteq \mathscr{V}(M \setminus \{x\}) \subsetneq M$ and consequently $\mathscr{V}(M \setminus \{x\}) = M \setminus \{x\}$, obtaining that $M \setminus \{x\}$ is a $\mathscr{V}$-monoid. $\square$

Recall that if S is a numerical semigroup other than $\mathbb{N}$, then $\mathrm{F}(S)$ becomes a minimal generator of $S \cup \{\mathrm{F}(S)\}$ greater than $\mathrm{F}(S \cup \{\mathrm{F}(S)\})$. The same holds with minimal $\mathscr{V}$-systems of generators.

Corollary 7.29. *Let S be a numerical semigroup. The following statements are equivalent.*

1) $S = T \cup \{\mathrm{F}(T)\}$ for some $T \in \mathscr{V}$.
2) $S \in \mathscr{V}$ and the minimal $\mathscr{V}$-system of generators of S contains an element greater than $\mathrm{F}(S)$.

Proof. 1) implies 2). As $T \in \mathscr{V}$ and since $\mathscr{V}$ is a Frobenius variety, $S = T \cup \{\mathrm{F}(T)\} \in \mathscr{V}$. By Proposition 7.28, we deduce that $\mathrm{F}(T)$ belongs to the minimal $\mathscr{V}$-system of generators of S. Furthermore, we clearly have that $\mathrm{F}(S) < \mathrm{F}(T)$.

2) implies 1). If $S \in \mathscr{V}$ and x is an element of the minimal $\mathscr{V}$-system of generators of S such that $\mathrm{F}(S) < x$, then by Proposition 7.28 we obtain that $T = S \setminus \{x\} \in \mathscr{V}$. We also have that $\mathrm{F}(T) = x$ and $S = T \cup \{x\}$. $\square$

Given a Frobenius variety $\mathscr{V}$, define $\mathscr{G}(\mathscr{V})$, the associated graph to $\mathscr{V}$, in the following way: the set of vertices of $\mathscr{G}(\mathscr{V})$ is $\mathscr{V}$ and $(T,S) \in \mathscr{V} \times \mathscr{V}$ is an edge of $\mathscr{G}(\mathscr{V})$ if and only if $S = T \cup \{\mathrm{F}(T)\}$.

Given $S \in \mathscr{V}(S)$, as $\mathrm{Ch}(S) \subseteq \mathscr{V}$, we deduce that there exists in $\mathscr{G}(\mathscr{V})$ a path connecting S with $\mathbb{N}$. By applying now Corollary 7.29 we obtain the following analogue to Proposition 7.1.

Theorem 7.30. *Let $\mathscr{V}$ be a Frobenius variety. The graph $\mathscr{G}(\mathscr{V})$ is a tree with root equal to $\mathbb{N}$. Furthermore, the sons of a vertex $S \in \mathscr{V}$ are $S \setminus \{x_1\}, \ldots, S \setminus \{x_r\}$ where $x_1, \ldots, x_r$ are the elements of the minimal $\mathscr{V}$-system of generators of S which are greater than $\mathrm{F}(S)$.*

In the following examples, for the sake of simplicity, we will write $\mathscr{V}(x_1, \ldots, x_n)$ instead of $\mathscr{V}(\{x_1, \ldots, x_n\})$.

Example 7.31 ([67]). Let $S_1 = \langle 5,7,9 \rangle = \{0,5,7,9,10,12,14,\rightarrow\}$ and $S_2 = \langle 4,6,7 \rangle = \{0,4,6,7,8,10,\rightarrow\}$ and let $\mathscr{V} = F(\{S_1,S_2\})$. Its associated tree is printed below.

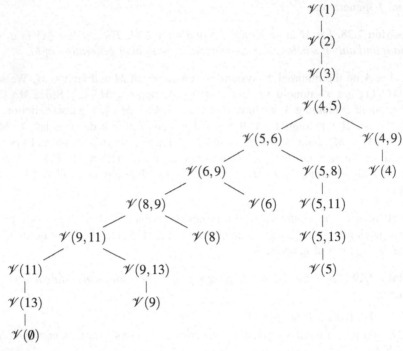

Example 7.32 ([88]). Let now $\mathscr{V}$ be the variety of Arf numerical semigroups. Part of the associated tree of the variety is drawn below (we indicate the Frobenius number of each of the nodes).

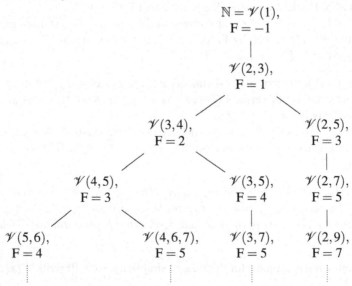

Each node has at most two sons (see [88]).

Example 7.33 ([89]). We now draw part of the tree associated to the Frobenius variety $\mathscr{V}$ of saturated numerical semigroups.

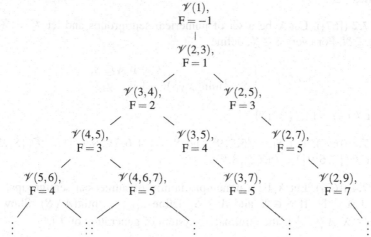

Every node has at most two sons, and there are no leaves (see [89]).

Example 7.34 ([11]). We now delineate part of the tree associated to the Frobenius variety $\mathscr{S}(x_1 + x_2 + x_3 - x_4)$ (see Exercise 7.7).

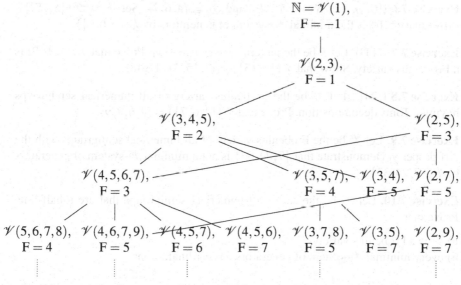

In this example there are elements with more than two sons. For instance, $\mathscr{V}(4,5,6,7)$ has four sons. The leaves in the portion of the directed acyclic graph drawn in the figure are $\mathscr{V}(3,4)$, $\mathscr{V}(4,5,6)$ and $\mathscr{V}(3,5)$.

Exercises

Exercise 7.1 ([67]). Prove that a Frobenius variety is finitely generated if and only if it is finite.

Exercise 7.2 ([67]). Let X be a set of numerical semigroups and let $\mathscr{V} = \mathscr{F}(X)$. Choose $A \subseteq \mathbb{N}$. For every $S \in X$, define

$$\mathscr{A}(S) = \begin{cases} S & \text{if } A \subseteq S, \\ S \cup \{\min(A \setminus S), \rightarrow\} & \text{if } A \nsubseteq S. \end{cases}$$

Show that $\mathscr{V}(A) = \bigcap_{S \in X} \mathscr{A}(S)$.

Exercise 7.3 ([67]). Let $S_1 = \langle 5,7,9 \rangle$ and $S_2 = \langle 4,6,7 \rangle$. Set $\mathscr{V} = \mathscr{F}(\{S_1, S_2\})$. Prove that $\mathscr{V}(\{5,6,9\}) = \langle 5,6,7,8,9 \rangle$.

Exercise 7.4 ([67]). Let X be a nonempty family of numerical semigroups, $\mathscr{V} = \mathscr{F}(X)$ and $A \subseteq \mathbb{N}$. If $S \in X$ and $A \nsubseteq S$, define $a_{(S,A)} = \min(A \setminus S)$. Show that $\{a_{(S,A)} \mid S \in X, A \nsubseteq S\}$ is the minimal $\mathscr{V}$-system of generators of $\mathscr{V}(A)$.

Exercise 7.5 ([67]). Let X be a nonempty family of numerical semigroups and let $\mathscr{V} = \mathscr{F}(X)$. Prove that every minimal $\mathscr{V}$-system of generators has cardinality less than or equal to X.

Exercise 7.6 ([67]). Let $S_1 = \langle 5,7,9 \rangle$ and $S_2 = \langle 4,6,7 \rangle$. Set $\mathscr{V} = \mathscr{F}(\{S_1, S_2\})$. Prove that $\{5,6\}$ is the minimal $\mathscr{V}$-system of generators of $\mathscr{V}(\{5,6,9\})$.

Exercise 7.7 ([11]). Let P be the pattern $x_1 + x_2 + x_3 - x_4$. Prove that $\mathscr{V} = \mathscr{S}(P)$ is a Frobenius variety. Show that $\mathscr{V}(\{7,15\}) = \langle 7,15,31,47,48 \rangle$.

Exercise 7.8 ([24]). Let $\mathscr{V}$ be the Frobenius variety of all numerical semigroups having a Toms decomposition. Prove that $\mathscr{V}(\{4,6,7\}) = \langle 4,6,7,9 \rangle$.

Exercise 7.9. Let $\mathscr{V}$ be the Frobenius variety of all numerical semigroups with the Arf property. Demonstrate that $\{7,24,33\}$ is not a minimal $\mathscr{V}$-system of generators of $\mathscr{V}(\{7,24,33\})$.

Exercise 7.10. Let $\mathscr{V}$ be the set of all numerical semigroups that are a half-line. Prove that

a) $\mathscr{V}$ is a Frobenius variety,
b) every minimal $\mathscr{V}$-system of generators has cardinality one.

Chapter 7
Presentations of a numerical semigroup

Introduction

Rédei in [53] shows that every congruence on $\mathbb{N}^n$ is finitely generated. This result has since been known as Rédei's theorem, and it is equivalent to the fact that every finitely generated (commutative) monoid is finitely presented. Rédei's proof is long and elaborated. Many other authors have given alternative and much simpler proofs than his (see for instance [31, 39, 41, 56]). Since numerical semigroups are cancellative monoids, a different approach can be chosen to prove Rédei's theorem. And this is precisely the path we choose in this chapter.

Another way to study presentations for finitely generated monoids relies on the following idea. Associated to a monoid M and to a ring R, one can define its semigroup ring $R[M]$ (see [37]). Presentations of monoids translate into binomial ideals on the ring of polynomials over R with as many unknowns as M has generators. Hilbert's basis theorem can be then used to derive Rédei's theorem (see [31, 41, 51]).

An important peculiarity of finitely generated cancellative (commutative) monoids is that minimal presentations with respect to set inclusion have minimal cardinality (see [93]). In this chapter we focus on the computation of a (and in fact all) minimal presentation of a numerical semigroup. The idea comes from Rosales' PhD thesis ([55]) and was published later in [59].

The cardinality of a minimal presentation of a numerical semigroup cannot be bounded in terms of its embedding dimension. This follows from Bresinsky's family of embedding dimension four numerical semigroups, which have arbitrarily large minimal presentations (see [14]). In this chapter we offer an upper bound for the cardinality of a minimal presentation in terms of the multiplicity of the semigroup.

J.C. Rosales, P.A. García-Sánchez, *Numerical Semigroups,*
Developments in Mathematics 20, DOI 10.1007/978-1-4419-0160-6_8,
© Springer Science+Business Media, LLC 2009

1 Free monoids and presentations

As in other branches of algebra, a monoid can be represented via a free object in the category of monoids modulo certain relations between the generators. To this end we need to introduce the concept of free monoid and state the isomorphism theorem for monoids.

Let X be a nonempty set. A *binary relation* on X is a subset σ of $X \times X$. If $(a, b) \in \sigma$, we write $x\sigma y$ and we say that x is σ-related with y. If σ is reflexive ($a\sigma a$ for all $a \in X$), symmetric ($a\sigma b$ implies $b\sigma a$) and transitive ($a\sigma b$ and $b\sigma c$ implies $a\sigma c$), then we say that σ is an *equivalence* binary relation. For every $a \in X$, we define its class modulo σ called its σ-class as

$$[a]_\sigma = \{b \in X \mid a\sigma b\}.$$

The set

$$\frac{X}{\sigma} = \{[a]_\sigma \mid a \in X\}$$

is the *quotient set* of X by σ, and it is a partition of X.

A *congruence* on a monoid M is an equivalence binary relation such that for all $a, b, c \in M$, if $a\sigma b$, then $(a + c)\sigma(b + c)$.

Lemma 8.1. *Let M be a monoid and σ a congruence on M. Then $\frac{M}{\sigma}$ is a monoid with the operation*

$$[a]_\sigma + [b]_\sigma = [a + b]_\sigma.$$

Proof. In order to see that $+$ is a map, we must prove that if $a\sigma b$ and $c\sigma d$, then $(a + c)\sigma(b + d)$. But this is an easy consequence of the fact that σ is a congruence, because $(a + c)\sigma(b + c)$ and $(b + c)\sigma(b + d)$, and by transitivity we obtain the desired property. The identity element is clearly $[0]_\sigma$, and the operation is trivially associative and commutative.   □

The monoid $(\frac{M}{\sigma}, +)$ is the *quotient monoid* of M modulo σ.

Let $f : X \to Y$ be a monoid homomorphism. We define the *kernel congruence* of f as

$$\ker(f) = \{(a, b) \in X \times X \mid f(a) = f(b)\}$$

(which clearly is a congruence on X because f is a homomorphism). The *image* of f, defined as

$$\operatorname{im}(f) = \{f(a) \mid a \in X\},$$

is a submonoid of Y.

Observe that there are slight differences between the usual definitions in Abelian groups and monoids. The condition $f(0) = 0$ for monoid morphisms is not imposed for group morphisms, since it follows directly from $f(0) = f(0 + 0) = f(0) + f(0)$. This is because in monoids we do not have a "cancellative" law (we will come back to this concept later). The other main difference is the definition of kernel. In group theory it is defined as the preimage of the identity element, that is, the set of elements that map to the identity element. This is because in group theory $f(a) = f(b)$ implies $f(a) - f(b) = 0$, or equivalently $f(a - b) = 0$. Thus in order to measure how far our morphism is from being injective, it suffices to study those elements that map to the identity element. This of course cannot be translated to monoids in general, since we do not have the notion of inverse. Although there are some differences, the first theorem of isomorphy still holds.

Proposition 8.2. *Let* $f : X \to Y$ *be a monoid homomorphism. Then*

$$\tilde{f} : \frac{X}{\ker(f)} \to \mathrm{im}(f), \; \tilde{f}([a]_{\ker(f)}) = f(a)$$

is a monoid isomorphism.

Proof. This definition of $\tilde{f}$ does not depend on the choice of the representative of $[a]_{\ker(f)}$, because of the meaning of $\ker(f)$. This map is both injective and surjective, and it is a homomorphism since f is a homomorphism.   □

The *free monoid* on a set X is defined as

$$\mathrm{Free}(X) = \{ \lambda_1 x_1 + \cdots + \lambda_k x_k \mid k \in \mathbb{N}, \lambda_1, \ldots, \lambda_k \in \mathbb{N}, x_1, \ldots, x_k \in X \}.$$

Addition is defined componentwise, that is,

$$(\lambda_1 x_1 + \cdots + \lambda_k x_k) + (\mu_1 x_1 + \cdots + \mu_k x_k) = (\lambda_1 + \mu_1) x_1 + \cdots + (\lambda_k + \mu_k) x_k$$

(some of the λ_i or μ_i can be zero). This set with this operation is a monoid. It coincides with the set of maps $\lambda : X \to \mathbb{N}$ with finite support ($\lambda(x) \neq 0$ for only finitely many $x \in X$). In order to simplify notation we will sometimes write $\mathrm{Free}(x_1, \ldots, x_n)$ to denote $\mathrm{Free}(\{x_1, \ldots, x_n\})$. Observe that $\mathrm{Free}(x_1, \ldots, x_n)$ (with $x_i \neq x_j$ for all $i \neq j$) is isomorphic to $\mathbb{N}^n$.

If M is generated by $\{m_1, \ldots, m_k\}$, then the map

$$\varphi : \mathrm{Free}(x_1, \ldots, x_k) \to M, \; \varphi(\lambda_1 x_1 + \cdots + \lambda_k x_k) = \lambda_1 m_1 + \cdots + \lambda_k m_k$$

is a monoid epimorphism (assuming that $x_i \neq x_j$ for $i \neq j$).

Note that we could have just defined φ as the (unique) monoid homomorphism determined by $\varphi(x_i) = m_i$ for all $i \in \{1, \ldots, k\}$. Hence by Proposition 8.2, M is isomorphic to $\mathrm{Free}(x_1, \ldots, x_n) / \ker(\varphi)$.

Proposition 8.3. *Let M be a monoid generated by $\{m_1, \ldots, m_k\}$. Let X be a set with cardinality k. Then there exists a congruence σ on $\mathrm{Free}(X)$ such that M is isomorphic to $\mathrm{Free}(X)/\sigma$.*

As a consequence of this property, every finitely generated monoid is isomorphic to $\mathbb{N}^k/\sigma$ for some positive k and some congruence σ on $\mathbb{N}^k$.

In order to achieve a finite representation of a finitely generated monoid, it still remains to see that the congruence σ can be "finitely described." We see next what we mean by this. First we introduce the concept of congruence generated by a set. We can do this because the intersection of congruences on a monoid is a congruence, and thus for a set X and a subset ρ of $\text{Free}(X) \times \text{Free}(X)$ we define the *congruence generated* by ρ as the intersection of all congruences on $\text{Free}(X)$ containing ρ. We denote this congruence by $\text{Cong}(\rho)$. We next see what are the elements in this set.

The binary relation $\Delta(X) = \{(x,x) \mid x \in X\}$ is known as the *diagonal* on X. For a binary relation ρ on a set X, its *inverse relation* is the set

$$\rho^{-1} = \{(b,a) \mid (a,b) \in \rho\}.$$

Proposition 8.4. *Let X be a nonempty set. Let $\rho \subseteq \text{Free}(X) \times \text{Free}(X)$. Define*

$$\rho^0 = \rho \cup \rho^{-1} \cup \Delta(\text{Free}(X)),$$

$$\rho^1 = \{(v+u, w+u) \mid (v,w) \in \rho^0, u \in \text{Free}(X)\}.$$

Then $\text{Cong}(\rho)$ *is the set of pairs* $(v,w) \in \text{Free}(X) \times \text{Free}(X)$ *such that there exist* $k \in \mathbb{N}$ *and* $v_0, \ldots, v_k \in \text{Free}(X)$ *with* $v_0 = v$, $v_k = w$ *and* $(v_i, v_{i+1}) \in \rho^1$ *for all* $0 \leq i \leq k-1$.

Proof. We first show that the set constructed in this way is a congruence. Let us call this set σ.

(1) Since $\Delta(\text{Free}(X)) \subseteq \sigma$, the binary relation σ is reflexive.
(2) If $(v,w) \in \sigma$, there exist $k \in \mathbb{N}$ and $v_0, \ldots, v_k \in \text{Free}(X)$ such that $v_0 = v$, $v_k = w$ and $(v_i, v_{i+1}) \in \rho^1$ for all $0 \leq i \leq k-1$. Since $(v_i, v_{i+1}) \in \rho^1$ implies that $(v_{i+1}, v_i) \in \rho^1$, defining $w_i = v_{k-i}$ for every $0 \leq i \leq k$, we obtain that $(w,v) \in \sigma$. Hence σ is symmetric.
(3) If (u,v) and (v,w) are in σ, then there exists $k, l \in \mathbb{N}$ and $v_0, \ldots, v_k, w_0, \ldots, w_l \in \text{Free}(X)$ such that $v_0 = u$, $v_k = w_0 = v$, $w_l = w$ and $(v_i, v_{i+1}), (w_j, w_{j+1}) \in \rho^1$ for all suitable i, j. By concatenating the sequences $\{v_i\}_i$ and $\{w_j\}_j$ we obtain $(u,w) \in \sigma$. Thus σ is transitive.
(4) Finally, let $(v,w) \in \sigma$ and $u \in \text{Free}(X)$. There exists $k \in \mathbb{N}$ and $v_0, \ldots, v_k \in \text{Free}(X)$ such that $v_0 = v$, $v_k = w$ and $(v_i, v_{i+1}) \in \rho^1$ for all $0 \leq i \leq k-1$. Defining $w_i = v_i + u$ for all $0 \leq i \leq k$ we have $(w_i, w_{i+1}) \in \rho^1$ and thus $(v+u, w+u) \in \sigma$.

It is clear that every congruence containing ρ must contain σ and this means that σ is the least congruence on $\text{Free}(X)$ that contains ρ, that is, $\sigma = \text{Cong}(\rho)$. $\qquad\square$

A congruence σ is *generated* by ρ if $\sigma = \text{Cong}(\rho)$. We say that ρ is a *system of generators* of σ. A congruence σ is *finitely generated* if there exists a system of generators of σ with finitely many elements.

A *presentation* of a finitely generated monoid M is a congruence on $\text{Free}(X)$, for some finite set X, such that $M \cong \text{Free}(X)/\text{Cong}(\rho)$. We say that M is a *finitely presented* monoid if ρ is finite.

Let X be a nonempty set and let σ be a congruence on $\text{Free}(X)$. An element (a,b) in $\sigma\{(0,0)\}$ is *irreducible* if it cannot be expressed as $(a,b) = (a_1, b_1) + (a_2, b_2)$

with $(a_1,b_1),(a_2,b_2) \in \sigma \setminus \{(0,0)\}$. Let $\mathrm{Irr}(\sigma)$ be the set of irreducible elements of σ. Note that σ is a submonoid of $\mathrm{Free}(X) \times \mathrm{Free}(X)$, since as we have seen in the proof of Lemma 8.1, if (a,b) and (c,d) are in σ, so is $(a+c,b+d)$, and by reflexivity $(0,0) \in \sigma$. As we see next, $\mathrm{Irr}(\sigma)$ generates σ as a monoid.

Proposition 8.5. *Let X be a nonempty finite set and let σ be a congruence on $\mathrm{Free}(X)$. Then $\sigma = \langle \mathrm{Irr}(\sigma) \rangle$.*

Proof. Assume that $X = \{x_1,\ldots,x_k\}$. We define the following order relation on $\mathrm{Free}(X)$: $\lambda_1 x_1 + \cdots + \lambda_k x_k \le \mu_1 x_1 + \cdots + \mu_k x_k$ if $\lambda_i \le \mu_i$ for all $i \in \{1,\ldots,k\}$. Given $x = \lambda_1 x_1 + \cdots + \lambda_k x_k \in \mathrm{Free}(X)$, the set of elements in $\mathrm{Free}(X)$ less than or equal to x is finite (there are $(\lambda_1 + 1) \cdots (\lambda_k + 1)$ of them). In the same way we can define on $\mathrm{Free}(X) \times \mathrm{Free}(X)$ the relation $(a,b) \le (c,d)$ if $a \le c$ and $b \le d$. We write $(a,b) < (c,d)$ if $(a,b) \le (c,d)$ and $(a,b) \ne (c,d)$. It follows that given (a,b) in $\mathrm{Free}(X) \times \mathrm{Free}(X)$ the set of elements $(c,d) \in \mathrm{Free}(X) \times \mathrm{Free}(X)$ such that $(c,d) \le (a,b)$ is finite.

Let $(a,b) \in \sigma$. If (a,b) is not irreducible, then $(a,b) = (a_1,b_1) + (a_1',b_1')$, with $(a_1,b_1),(a_1',b_1') \in \sigma \setminus \{(0,0)\}$. Then $(a_1,b_1) < (a,b)$ and $(a_1',b_1') < (a,b)$. If any of them is not irreducible, then we can repeat the process with them. In this way we can construct a binary tree rooted in (a,b) and every node is the sum of its two sons. A leaf in the tree is an irreducible element of σ. Since all the elements in the tree are smaller than (a,b) (except (a,b) itself), and the number of elements in $\mathrm{Free}(X) \times \mathrm{Free}(X)$ smaller than (a,b) is finite, this tree has only finitely many nodes and thus finitely many leaves. From the way the tree is constructed, (a,b) is the sum of all the leaves in the tree, obtaining in this way that $(a,b) \in \langle \mathrm{Irr}(\sigma) \rangle$. $\square$

A monoid M is *cancellative* if for any $a,b,c \in M$, $a+c = b+c$ implies $a = b$. We say that a congruence σ on $\mathrm{Free}(X)$ is *cancellative* if $\mathrm{Free}(X)/\sigma$ is cancellative. We now see that, as a consequence of Dickson's lemma, finitely generated cancellative monoids are always finitely presented.

Lemma 8.6 (Dickson's lemma). *Let X be a nonempty set with finitely many elements. Let N be a nonempty subset of $\mathrm{Free}(X)$. The set $M = \mathrm{Minimals}_{\le}(N)$ has finitely many elements.*

Proof. Assume that $X = \{x_1,\ldots,x_n\}$. We use induction on n. For $n = 1$, the result follows easily from the fact that $\le$ is a well order on $\mathrm{Free}(x_1) = \{\lambda x_1 \mid \lambda \in \mathbb{N}\}$, because $\le$ is a well order on $\mathbb{N}$ (every subset of $\mathbb{N}$ has a minimum).

Assume that the statement is true for $n-1$ and let us show it for n. Choose an element $a_1 x_1 + \cdots + a_n x_n \in M$. For each $1 \le i \le n$ and each $0 \le j \le a_i$ define

$$M_{ij} = \{(m_1 x_1 + \cdots + m_n x_n) \in M \mid m_i = j\}$$

and

$$B_{ij} = \{b_1 x_1 + \cdots + b_{n-1} x_{n-1} \in \mathrm{Free}(x_1,\ldots,x_{n-1}) \mid$$
$$b_1 x_1 + \cdots + b_{i-1} x_{i-1} + j x_i + b_i x_{i+1} + \cdots + b_{n-1} x_n \in M_{ij}\}.$$

Observe that Minimals$_\leq(M) = M$ and for this reason Minimals$_\leq(M_{ij}) = M_{ij}$ and
Minimals$_\leq(B_{ij}) = B_{ij}$. By induction hypothesis, B_{ij} must be finite. Hence, M_{ij} is
finite as well. Since there are finitely many sets M_{ij}, the set $\bigcup M_{ij}$ is again finite
and nonempty. Hence it is enough to show that $M \subseteq \bigcup M_{ij}$. Take $m_1x_1 + \cdots + m_nx_n$
to be an element in M. There exists $i \in \{1, \ldots, n\}$ such that $m_i \leq a_i$ (if this were
not the case, $a_1x_1 + \cdots + a_nx_2 < m_1x_1 + \cdots + m_nx_n$ and this is impossible, since
$m_1x_1 + \cdots + m_nx_n$ is a minimal element of N). Hence $m_1x_1 + \cdots + x_nm_n \in M_{im_i}$. $\square$

Given $a = a_1x_1 + \cdots + a_nx_n, b = b_1x_1 + \cdots + b_nx_n \in \text{Free}(x_1, \ldots, x_n)$. If $a \leq b$,
then $a_i \leq b_i$ for all $i \in \{1, \ldots, n\}$. Thus $(b_1 - a_1)x_1 + \cdots + (b_n - a_n)x_n \in \text{Free}(X)$.
We denote this element by $b - a$.

Proposition 8.7. *Let X be a nonempty set with finitely many elements. Let σ be a
cancellative congruence on $\text{Free}(X)$. Then the set $\text{Irr}(\sigma)$ is finite.*

Proof. If we prove that

$$\text{Irr}(\sigma) \subseteq \text{Minimals}_\leq(\sigma \setminus \{0,0\}),$$

then by using Dickson's lemma the set $\text{Irr}(\sigma)$ must be finite. Let (a,b) be an element in $\text{Irr}(\sigma)$. Assume that (a,b) is not a minimal element of $\sigma \setminus \{(0,0)\}$. Thus
there exists $(a_1,b_1) \in \text{Minimals}_\leq(\sigma \setminus \{0,0\})$ such that $(a_1,b_1) < (a,b)$. Hence
$(a,b) = (a_1,b_1) + (a_2,b_2)$ with $(a_2,b_2) = (a - a_1, b - b_1) \in \text{Free}(X) \times \text{Free}(X)$.
Since $(a,b) = (a_1 + a_2, b_1 + b_2) \in \sigma$ and $(b_1,a_1) \in \sigma$, we have that $(a_2 + (a_1 +
b_1), b_2 + (a_1 + b_1)) \in \sigma$. By hypothesis, $\text{Free}(X)/\sigma$ is cancellative and this implies
that from the equality

$$[a_2 + (a_1 + b_1)]_\sigma = [a_2]_\sigma + [a_1 + b_1]_\sigma = [b_2]_\sigma + [a_1 + b_1]_\sigma = [b_2 + (a_1 + b_1)]_\sigma$$

we obtain that $[a_2]_\sigma = [b_2]_\sigma$. Thus $(a_2,b_2) \in \sigma$, which means that $(a,b) \notin \text{Irr}(\sigma)$.
The contradiction comes from the assumption $(a,b) \notin \text{Minimals}_\leq(\sigma \setminus \{0,0\})$. $\square$

With this we obtain the desired consequence.

Corollary 8.8. *Every finitely generated cancellative monoid is finitely presented.*

2 Minimal presentations of a numerical semigroup

Every numerical semigroup is a finitely generated cancellative monoid and thus it is
finitely presented. We characterize in this section those presentations of numerical
semigroups that are minimal. We will see that the concepts of minimal with respect
to set inclusion and cardinality coincide for numerical semigroups.

Let σ be a congruence on $\text{Free}(x_1, \ldots, x_n)$ and let ρ be a system of generators of
σ. We say that ρ is a *minimal relation* if the cardinality of ρ is the least possible
among the cardinalities of systems of generators of σ. Let S be a numerical semigroup minimally generated by $\{n_1, \ldots, n_e\}$ and let $X = \{x_1, \ldots, x_e\}$ with $x_i \neq x_j$ for

all $i \neq j$. We say that ρ is a *minimal presentation* if ρ is a minimal relation of the kernel congruence of $\varphi : \mathrm{Free}(x_1,\ldots,x_e) \to S$, $\varphi(a_1x_1 + \cdots + a_ex_e) = a_1n_1 + \cdots + a_en_e$. In this section, we use σ to denote the kernel congruence of φ.

Given $n \in \mathbb{N}$ the *set of expressions* of n in S is defined as

$$ Z(n) = \varphi^{-1}(n) = \{ a_1x_1 + \cdots + a_ex_e \mid a_1n_1 + \cdots + a_en_e = n \}. $$

For $a = a_1x_1 + \cdots + a_ex_e, b = b_1x_1 + \cdots + b_ex_e \in \mathrm{Free}(x_1,\ldots,x_e)$. Define the *dot product* of a and b as

$$ a \cdot b = a_1b_1 + \cdots + a_eb_e. $$

We define the following relation on $\mathrm{Free}(x_1,\ldots,x_e)$. For $a, b \in \mathrm{Free}(x_1,\ldots,x_e)$, $a\mathrm{R}b$ if either $a = b = 0$ or there exist $k_1,\ldots,k_l \in Z(n)$ for some $n \in S$ such that $k_1 = a$, $k_l = b$ and $k_i \cdot k_{i+1} \neq 0$ for all $i \in \{1,\ldots,l-1\}$. This is an equivalence binary relation on $\mathrm{Free}(X)$. The elements of $\mathrm{Free}(X)/\mathrm{R}$ are called R-*classes*.

The concept of graph differs slightly from that of directed graph (see page 91). We recall it now since it is of crucial importance in what follows. A (nondirected) *graph* G is a pair (V,E) where V is a set whose elements are known as the *vertices* of G and E is a subset of $\{ \{u,v\} \mid u,v \in V, u \neq v \}$. The unordered pair $\{u,v\}$ will be denoted as $\overline{uv}$, and if it belongs to E, then we say that it is an *edge* of G.

A sequence of edges of the form $\overline{v_0v_1}, \overline{v_1v_2}, \ldots, \overline{v_{m-1}v_m}$ is known as a *path* of length m connecting v_0 and v_m. A graph is *connected* if for any two vertices of the graph, there is a path connecting them. It is well known that a connected graph with n vertices has at least $n-1$ edges (see [48]). A *tree* is a connected graph with n vertices and $n-1$ edges for some positive integer n (this is one of the many characterizations of a tree). A subgraph of the graph $G = (V,E)$ is a graph $G' = (V',E')$ such that $V' \subseteq V$ and $E' \subseteq E$. It is also well known that any connected graph G with n vertices has a subgraph with the same vertices that is a tree. This tree is called the *generating tree* of G.

Let X be a nonempty set, let $P = \{X_1,\ldots,X_r\}$ be a partition of X and let γ be a binary relation on X. The graph associated to γ with respect to the partition P is $G_\gamma = (V,E)$, where $V = P$ and $\overline{X_iX_j} \in E$ with $i \neq j$ if there exists $x \in X_i$ and $y \in X_j$ such that $(x,y) \in \gamma \cup \gamma^{-1}$.

Let $n \in \mathbb{N}$ and let $X_1,\ldots,X_r$ be the R-classes contained in $Z(n)$. If β is a binary relation on $\mathrm{Free}(x_1,\ldots,x_e)$, we denote by

$$ \beta_n = \beta \cap (Z(n) \times Z(n)) $$

and by G_{β_n} the graph associated to the partition $\{X_1,\ldots,X_r\}$ of $Z(n)$. As we see next, these graphs are crucial for our characterization of minimal presentations of a numerical semigroup.

Lemma 8.9. *Let $n \in \mathbb{N}$. If β is a binary relation on $\mathrm{Free}(x_1,\ldots,x_e)$ generating σ, then G_{β_n} is connected.*

Proof. Let $X_1,\ldots,X_r$ be the R-classes contained in $Z(n)$. Let t and s be in $\{1,\ldots,r\}$ with $t \neq s$. We prove that there exists a path connecting X_t and X_s. Let $k \in X_t$ and

$h \in X_s$. As $k, h \in Z(n)$, $k\sigma h$. Since $\mathrm{Cong}(\beta) = \sigma$, by Proposition 8.4, there exist $b_0, b_1, \ldots, b_l \in \mathrm{Free}(x_1, \ldots, x_e)$, such that $k = b_0$, $h = b_l$ and $(b_i, b_{i+1}) \in \beta^1$ for $i \in \{0, \ldots, l-1\}$. Hence there exist for all $i \in \{0, \ldots, l-1\}$, $z_i \in \mathrm{Free}(x_1, \ldots, x_e)$ and $(x_i, y_i) \in \beta \cup \beta^{-1}$ such that $(b_i, b_{i+1}) = (x_i + z_i, y_i + z_i)$. If $z_i \neq 0$, then $b_i \mathrm{R} b_{i+1}$. And if $z_i = 0$, then $\{b_i, b_{i+1}\} \subseteq \beta_n$. Hence the nonordered pairs $\{b_i, b_{i+1}\}$ such that $(b_i, b_{i+1}) \notin \mathrm{R}$ yield the path in G_{β_n} connecting X_t with X_s. $\square$

Theorem 8.10. *Let β be a binary relation on $\mathrm{Free}(x_1, \ldots, x_e)$ with $x_i \neq x_j$ for $i \neq j$. Then $\mathrm{Cong}(\beta) = \sigma$ if and only if G_{β_n} is connected for all $n \in \mathbb{N}$.*

Proof. The necessary condition is the preceding lemma. So let us focus on the sufficiency. Observe that in order to prove that $\mathrm{Cong}(\beta) = \sigma$, it suffices to show that for all $n \in \mathbb{N}$, if $k, k' \in Z(n)$, then $(k, k') \in \mathrm{Cong}(\beta)$. We use induction on n. If $n = 0$, then $Z(n) = \{0\}$ and trivially $(0, 0) \in \mathrm{Cong}(\beta)$. Hence assume as induction hypothesis that the result holds for integers less than n. If $n \notin S$, then $Z(n)$ is empty and the result follows by vacuity. Thus we can assume that $n \in S$. We distinguish two cases.

1) If $k\mathrm{R}k'$, then there exist $k_0, \ldots, k_l \in Z(n)$ such that $k = k_0$, $k' = k_l$ and for all $i \in \{0, \ldots, l-1\}$, $k_i \cdot k_{i+1} \neq 0$, or equivalently, there exists $t \in \{1, \ldots, e\}$ such that $x_t \leq k_i$ and $x_t \leq k_{i+1}$. It follows that $k_i - x_t, k_{i+1} - x_t \in Z(n - n_t)$. By induction hypothesis $(k_i - x_t, k_{i+1} - x_t) \in \mathrm{Cong}(\beta)$. As $\mathrm{Cong}(\beta)$ is a congruence, $(k_i, k_{i+1}) \in \mathrm{Cong}(\beta)$ and by transitivity, this leads to $(k, k') \in \mathrm{Cong}(\beta)$.

2) If $(k, k') \notin \mathrm{R}$, then there are at most two R-classes on $Z(n)$. Assume that the sequence $X_1, \ldots, X_r$ defines a path connecting X_1 and X_r, where X_1 is the R-class of k and X_r that of k'. For every $i \in \{1, \ldots, r-1\}$, as $\overline{X_i X_{i+1}}$ is an edge of G_{β_n}, there exists $\{a_i, b_i\} \subseteq \beta_n$ with $a_i \in X_i$ and $b_i \in X_{i+1}$. Hence $(a_i, b_i) \in \mathrm{Cong}(\beta)$ for all i. Moreover, by 1), both (k, a_1) and (b_{r-1}, k') belong to $\mathrm{Cong}(\beta)$. By transitivity, we conclude that $(k, k') \in \mathrm{Cong}(\beta)$. $\square$

As we already know what are the smallest connected subgraphs of a graph with the same vertices, we can find a characterization for minimal presentations of a numerical semigroup.

Corollary 8.11. *Let S be a numerical semigroup. A subset β of σ is a minimal presentation of S if and only if the cardinality of β_n equals the number of R-classes in $Z(n)$ minus one and G_{β_n} is connected for all $n \in S$.*

Remark 8.12. Observe that if β is a minimal presentation for S, then G_{β_n} is a generating tree of G_{σ_n} for all $n \in \mathbb{N}$.

In order to obtain a minimal relation for σ (equivalently, a minimal presentation for S), it suffices to focus on those $n \in S$ for which the number of R-classes in $Z(n)$ is greater than or equal to two. If $X_1, \ldots, X_r$ are these R-classes, we construct a tree T_n whose vertices are $X_1, \ldots, X_r$. For every $\overline{X_i X_j}$ edge of T_n, take $a_i \in X_i$ and $b_i \in X_j$. Define β_n as the set of all (a_i, b_i) built in this way. Set $\beta_n = \emptyset$ if there are less than two R-classes in $Z(n)$. Then $\beta = \bigcup_{n \in S} \beta_n$ is a minimal relation of σ.

Recall that we have defined a minimal relation of σ as a generating system with minimal cardinality. The above result also implies that the concept of minimal generating system of σ with respect to cardinality and inclusion coincide. This does not hold for monoids in general (monoids for which these two concepts are the same are studied in [93]).

Corollary 8.13. *The concept of minimal presentation with respect to cardinality and set inclusion coincide for any numerical semigroup. In particular, all minimal presentations have the same cardinality.*

3 Computing minimal presentations

We present a method described in [59] to compute a minimal presentation of a numerical semigroup. In this section, $S = \langle n_1, \ldots, n_e \rangle$ and σ are as in the preceding section.

For every $n \in S$, define the graph $G_n = (V_n, E_n)$ with

$$V_n = \{ n_i \mid n - n_i \in S \}$$

and

$$E_n = \{ \overline{n_i n_j} \mid n - (n_i + n_j) \in S, i \neq j \}.$$

We will refer to this graph as the *graph associated* to n in S. We are going to relate these graphs with the ones appearing in the last section.

A *connected component* of a graph G is a maximal connected subgraph of G. If G is connected, then it has only a connected component. We next describe the connected components of the graph associated to an element in a numerical semigroup.

Let n be an element of S. If $X_1, \ldots, X_r$ are the R-classes of $Z(n)$, for all $i \in \{1, \ldots, r\}$ define

$$A_i = \{ n_j \mid x_j \leq x \text{ for some } x \in X_i \}.$$

These sets contain the set of vertices of the different connected components of G_n. To prove this, we first must show that $\{A_1, \ldots, A_r\}$ is a partition of V_n.

Lemma 8.14. *The set $\{A_1, \ldots, A_r\}$ is a partition of V_n.*

Proof. If $n_j \in V_n$, then $n - n_j \in S$. Take $a \in Z(n - n_j)$. Then $a + x_j \in Z(n)$ and thus there exists $i \in \{1, \ldots, r\}$ such that $a + x_j \in X_i$. Hence $n_j \in A_i$. This proves that $V_n \subseteq A_1 \cup \cdots \cup A_r$, and consequently $V_n = A_1 \cup \cdots \cup A_r$.

Now assume that $n_k \in A_i \cap A_j$ with $i \neq j$. Then there exists $x \in X_i$ and $y \in X_j$ such that $x_k \leq x$ and $x_k \leq y$. This implies that $x \cdot y \neq 0$ and consequently $x R y$. But this is impossible because X_i and X_j are different R-classes. This proves that $\{A_1, \ldots, A_r\}$ is a partition of V_n. □

Now we show that there is no edge in G_n connecting a vertex in A_i with a vertex in A_j when $i \neq j$.

Lemma 8.15. *If $n_k \in A_i$ and $n_l \in A_j$ with $i \neq j$, then $\overline{n_k n_l} \notin E_n$.*

Proof. Observe that by Lemma 8.14, k cannot be equal to l. Assume to the contrary that $\overline{n_k n_l} \in E_n$. Then $n - (n_k + n_l) \in S$. Take $a \in Z(n - (n_k + n_l))$. Then $a + x_k + x_l \in Z(n)$. From $n_k \in A_i$ and $n_l \in A_j$ we deduce that there exists $y \in X_i$ with $x_k \leq y$ and $z \in X_j$ with $x_l \leq z$. This leads to $y \cdot (a + x_k + x_l) \neq 0$ and $(a + x_k + x_l) \cdot z \neq 0$, which implies that $y \mathrm{R} z$, contradicting that X_i and X_j are different R-classes. □

And finally we prove that any two vertices in the same A_s are connected with a path in G_n.

Lemma 8.16. *If $n_k, n_l \in A_s$, with $l \neq k$, then there is a path in G_n with vertices in A_s joining n_k with n_l.*

Proof. By definition, there exist $a, b \in X_s$ such that $x_l \leq a$ and $x_k \leq b$. As $a \mathrm{R} b$, there exist a sequence $a_0, \ldots, a_r$ of elements in X_s such that $a_0 = a$, $a_r = b$ and $a_i \cdot a_{i+1} \neq 0$ for all $i \in \{0, \ldots, r-1\}$. We proceed by induction on r.

If $r = 1$ (the case $r = 0$ makes no sense since $k \neq l$), then $a \cdot b \neq 0$. Thus there exists $m \in \{1, \ldots, e\}$ such that $x_m \leq a$ and $x_m \leq b$. It follows that $x_l + x_m \leq a$ and $x_m + x_k \leq b$. And this implies that $n - (n_l + n_m) \in S$ and $n - (n_m + n_k) \in S$. This gives the path joining n_l and n_k (note that it may happen that either $n_m = n_l$ or $n_m = n_k$, but not both). Observe that from the definition of A_s, n_l, n_m and n_k belong to A_s.

Now assume that $r > 1$ and as induction hypothesis that the result holds for all sequences of length less than r. As $k \neq l$, there exists $m \in \{0, \ldots, r-1\}$ and $j \neq l$ such that $x_l \leq a_m$, $x_j \leq a_m$, $x_j \leq a_{m+1}$ and $x_l \nleq a_{m+1}$. Choose m to be minimum fulfilling this condition. From $x_l \leq a_m$ and $x_j \leq a_m$, we deduce that $\overline{n_l n_j} \in E_n$, and as $a_m \in X_s$, both n_l and n_l are in A_s. If $j = k$, then we are done. Otherwise, the induction hypothesis applied to $a_{m+1}, \ldots, a_r$ ensures the existence of a path connecting n_j and n_k in G_n with vertices in A_s. Adding at the beginning of this path the edge $\overline{n_l n_j}$, we conclude the proof. □

With all this information it is easy to see that the sets of vertices of the different connected components of G_n are $A_1, \ldots, A_r$. Thus the number of connected components of G_n equals the number of R-classes in $Z(n)$.

Theorem 8.17. *Let S be a numerical semigroup and let n be a nonzero element of S. The number of connected components of G_n equals the number of R-classes in $Z(n)$.*

Proof. Let $X_1, \ldots, X_r$ be the different R-classes in $Z(n)$. For every $i \in \{1, \ldots, r\}$, define the subgraph $G_n^i = (V_n^i, E_n^i)$ of G_n as follows. Set $V_n^i = A_i$ and $E_n^i = \{\overline{n_i n_i} \in E_n \mid n_i, n_j \in A_i\}$. From Lemmas 8.14, 8.15 and 8.16 we deduce that these are the connected components of G. □

Remark 8.18. From these results we deduce that to obtain a (in fact any) minimal presentation of S, we only have to find those $n \in S$ for which G_n is not connected. Let n be such that G_n is not connected and let $G_n^1, \ldots, G_n^r$ be its connected components. Choose now for every $i \in \{1, \ldots, r\}$ a vertex $n_{k_i} \in V_n^i$ and an element $\alpha_i \in Z(n)$ such that $x_{k_i} \leq \alpha_i$. Then $[\alpha_1]_{\mathrm{R}}, \ldots, [\alpha_r]_{\mathrm{R}}$ are the different R-classes of $Z(n)$. If we define $\rho_n = \{(\alpha_2, \alpha_1), \ldots, (\alpha_r, \alpha_1)\}$, then G_{ρ_n} is a tree. Hence, by Corollary 8.11, ρ is a minimal presentation of S.

From this we deduce that if we want to compute a minimal presentation by using this idea, we must find the elements in a numerical semigroup whose associated graph is not connected. The following property makes this search possible.

Proposition 8.19. *If G_n is not connected, then $n = w + n_j$ with $w \in \mathrm{Ap}(S, n_1) \setminus \{0\}$ and $j \in \{2, \ldots, e\}$.*

Proof. If $n - n_1 \notin S$, then $n = w \in \mathrm{Ap}(S, n_1) \setminus \{0\}$. Since G_n is not connected, there exist $n_i, n_j \in V_n$ with $i \neq j \geq 2$ such that $\overline{n_i n_j} \notin E_n$. This implies that $n = (n - n_j) + n_j$, and clearly $n - n_j \in \mathrm{Ap}(S, n_1) \setminus \{0\}$ ($n \neq n_j$ because $n - n_i \in S$ and $\{n_1, \ldots, n_e\}$ is a minimal generating system of S).

Now assume that $n - n_1 \in S$. As G_n is not connected, there exists $n_i \in V_n$ with $i > 1$, such that $n - (n_1 + n_i) \notin S$ ($\overline{n_1 n_i} \notin E_n$). As $n - n_i \in S$ and $n - n_i - n_1 \notin S$, we have that $n - n_i \in \mathrm{Ap}(S, n_1) \setminus \{0\}$ (again $n - n_i \neq 0$ because $n - n_1 \in S$). We can write $n = (n - n_i) + n_i$. $\qquad \square$

Remark 8.20. Note that this proposition gives a finite set in which are contained the elements $n \in S$ such that G_n is not connected. If we apply to this set the procedure described in Remark 8.18, we have an algorithmic procedure to compute a minimal presentation of a numerical semigroup.

Example 8.21 ([80]). Let
$$S = \langle 5, 7, 9, 11 \rangle.$$

As
$$S = \{0, 5, 7, 9, 10, 11, 12, 14, 15, 16, 17, 18, \ldots\},$$
$$\mathrm{Ap}(S, 5) = \{0, 7, 9, 11, 18\}.$$

Now we calculate $(\mathrm{Ap}(S, 5) \setminus \{0\}) + \{7, 9, 11\}$ obtaining

$$\{14, 16, 18, 20, 22, 25, 27, 29\}.$$

Hence, we look at the graphs G_n, with n in the set $(\mathrm{Ap}(S, 5) \setminus \{0\}) + \{7, 9, 11\}$.

We find an expression of 14 in which 5 is involved: $14 = 1 \times 5 + 1 \times 9$; and another in which 7 is used $14 = 2 \times 7$. We proceed analogously with 16, 18, 20 and 22.

Graph	Connected components	Relations
G_{14}	$\{5, 9\}, \{7\}$	$(x_1 + x_3, 2x_2)$
G_{16}	$\{5, 11\}, \{7, 9\}$	$(x_1 + x_4, x_2 + x_3)$
G_{18}	$\{7, 11\}, \{9\}$	$(x_2 + x_4, 2x_3)$
G_{20}	$\{5\}, \{9, 11\}$	$(4x_1, x_3 + x_4)$
G_{22}	$\{5, 7\}, \{11\}$	$(3x_1 + x_2, 2x_4)$
G_{25}	$\{5, 7, 9, 11\}$	
G_{27}	$\{5, 7, 9, 11\}$	
G_{29}	$\{5, 7, 9, 11\}$	

Therefore a minimal presentation of S is

$$\rho = \{(x_1 + x_3, 2x_2), (x_1 + x_4, x_2 + x_3), (x_2 + x_4, 2x_3), (4x_1, x_3 + x_4), (3x_1 + x_2, 2x_4)\}.$$

Example 8.22. Let $S = \langle n_1, n_2 \rangle$ with $\gcd\{n_1, n_2\} = 1$. Then a minimal presentation for S is $\{(n_2 x_1, n_1 x_2)\}$. The Apéry set of n_1 in S is

$$\mathrm{Ap}(S, n_1) = \{0, n_2, 2n_2, \ldots, (n_1 - 1)n_2\}.$$

The only n for which G_n is not connected is $n = n_1 n_2 = (n_1 - 1)n_2 + n_2$.

Example 8.23. Assume now that S is minimally generated by $\{n_1, n_2, n_3\}$. For every $r \in \{1, 2, 3\}$, define

$$c_r = \min\{k \in \mathbb{N} \setminus \{0\} \mid kn_r \in \langle n_s, n_t \rangle, \{r, s, t\} = \{1, 2, 3\}\}.$$

For $n \in \{c_1 n_1, c_2 n_2, c_3 n_3\}$, the graph G_n is not connected, and these are the only elements in S fulfilling this condition. We distinguish three cases.

1) If $c_1 n_1 = c_2 n_2 = c_3 n_3$, then

$$\{(c_1 x_1, c_2 x_2), (c_1 x_1, c_3 x_3)\}$$

is a minimal presentation for S.
2) Assume now that $c_1 n_1 \neq c_2 n_2 = c_3 n_3$ (we omit the other similar cases). If $c_1 n_1 = \lambda n_2 + \mu n_3$ with $\lambda, \mu \in N$, then

$$\{(c_1 x_1, \lambda x_2 + \mu x_3), (c_2 x_2, c_3 x_3)\}$$

is a minimal presentation for S.
3) If the cardinality of $\{c_1 n_1, c_2 n_2, c_3 n_3\}$ is three, then suppose that

$$c_1 n_1 = r_{12} n_2 + r_{13} n_3,$$
$$c_2 n_2 = r_{21} n_1 + r_{23} n_3,$$
$$c_3 n_3 = r_{31} n_1 + r_{32} n_2,$$

for some nonnegative integers r_{ij}. Then

$$\{(c_1 x_1, r_{12} x_2 + r_{13} x_3), (c_2 x_2, r_{21} x_1 + r_{23} x_3), (c_3 x_3, r_{31} x_1 + r_{32} x_2)\}$$

is a minimal presentation of S.

Note that $\langle 6, 10, 15 \rangle$, $\langle 4, 6, 7 \rangle$ and $\langle 5, 6, 7 \rangle$ are examples of 1), 2) and 3), respectively. We show how we can compute the presentations of these examples by using GAP.

```
gap> MinimalPresentationOfNumericalSemigroup(
>NumericalSemigroup(6,10,15));
[ [ [ 5, 0, 0 ], [ 0, 3, 0 ] ], [ [ 5, 0, 0 ],
[ 0, 0, 2 ] ] ]
```

```
gap> GraphAssociatedToElementInNumericalSemigroup(30,
> NumericalSemigroup(6,10,15));
[ [ 6, 10, 15 ], [ [ ] ] ]

gap> MinimalPresentationOfNumericalSemigroup(
>NumericalSemigroup(4,6,7));
[ [ [ 3, 0, 0 ], [ 0, 2, 0 ] ], [ [ 2, 1, 0 ],
[ 0, 0, 2 ] ] ]

gap> GraphAssociatedToElementInNumericalSemigroup(12,
> NumericalSemigroup(4,6,7));
[ [ 4, 6 ], [ [ ] ] ]

gap> MinimalPresentationOfNumericalSemigroup(
>NumericalSemigroup(5,6,7));
[ [ [ 1, 0, 1 ], [ 0, 2, 0 ] ], [ [ 4, 0, 0 ],
[ 0, 1, 2 ] ], [ [ 3, 1, 0 ], [ 0, 0, 3 ] ] ]

gap> GraphAssociatedToElementInNumericalSemigroup(12,
> NumericalSemigroup(5,6,7));
[ [ 5, 6, 7 ], [ [ 5, 7 ] ] ]
```

The output of the minimal presentation command is a list of pairs, each one containing a list of coefficients, by using the isomorphism $\mathbb{N}^3 \cong \mathrm{Free}(x_1,x_2,x_3)$. As for the graphs, the output is a list whose first element is the list of vertices and the second the list of edges of the graph.

4 An upper bound for the cardinality of a minimal presentation

We assume, unless otherwise stated, that S is minimally generated by $\{n_1,\ldots,n_e\}$. In this section we are concerned with upper bounding the cardinality of a minimal presentation for S. In order to find this bound we must deepen in the study of those elements n in S for which G_n is not connected. The bound given here is sharp, in the sense that it is reached by a large family of numerical semigroups: those with maximal embedding dimension (however this does not mean that better bounds can be given; see the exercises at the end of this chapter and the references given there). In fact, this characterizes numerical semigroups with maximal embedding dimension.

Lemma 8.24. *Let n be an element in S with nonconnected G_n. Then there exists $n_k \in \{n_2,\ldots,n_e\}$ and $w \in \mathrm{Ap}(S,n_1) \cap \cdots \cap \mathrm{Ap}(S,n_{k-1})$ such that*

1) $n = w + n_k \notin \mathrm{Ap}(S,n_1) \cap \cdots \cap \mathrm{Ap}(S,n_{k-1})$,
2) $w' + n_k \in \mathrm{Ap}(S,n_1) \cap \cdots \cap \mathrm{Ap}(S,n_{k-1})$, for all $w' \in S \setminus \{w\}$ such that $w' \leq_S w$.

Proof. Let $i = \min\{j \in \{1,\ldots,e\} \mid n_j \in V_n\}$. As G_n is not connected, we can also find k the minimum of $j \in \{1,\ldots,e\}$ such that n_i and n_j are in different connected components of G_n.

1) Since $n_k \in V_n$, $w = n - n_k \in S$. Assume that $w - n_j \in S$ for some $j \in \{1,\ldots,k-1\}$. Then $\overline{n_j n_k} \in E_n$ and thus n_j and n_k are in the same connected component. This in particular means that n_j and n_i are in different connected components. The minimality of k ensures that $k \leq j$, a contradiction. Hence $n = w + n_k$ with $w \in \mathrm{Ap}(S,n_1) \cap \cdots \cap \mathrm{Ap}(S,n_{k-1})$. As $i < k$, we have that $n \notin \mathrm{Ap}(S,n_1) \cap \cdots \cap \mathrm{Ap}(S,n_{k-1})$.

2) As $w - w' \in S \setminus \{0\}$, there exists $l \in \{1,\ldots,e\}$ such that $w - w' - n_l \in S$. Hence $n - (n_k + n_l) \in S$, which implies that n_k and n_l are in the same connected component of G_n. If $w' + n_k \notin \mathrm{Ap}(S,n_1) \cap \cdots \cap \mathrm{Ap}(S,n_{k-1})$, then there exists $j < k$ such that $w' + n_k - n_j \in S$. This leads to $(w' + n_k) - n_j + (w - w' - n_l) = n - (n_j + n_l) \in S$, which implies that n_j and n_k are in the same connected component of n_k, contradicting the minimality of k. □

For $k \in \{2,\ldots,e\}$, define D_k as the set of $w \in \mathrm{Ap}(S,n_1) \cap \cdots \cap \mathrm{Ap}(S,n_{k-1})$ fulfilling Conditions 1) and 2) of Lemma 8.24.

Lemma 8.25. *The cardinality of any minimal presentation of S is less than or equal to the sum of the cardinalities of the sets $D_2,\ldots,D_e$.*

Proof. Denote by $D = \bigcup_{k=1}^{e} D_k$ the disjoint union of the sets D_k. We will refer to its elements as (w,k) pointing out in this way that $w \in D_k$. Let ρ be a minimal presentation of S such that if $(a,b) \in \rho$, $n = \varphi(a) = \varphi(b)$ and $i = \min\{j \in \{1,\ldots,e\} \mid n_j \in V_n\}$, then $x_i \leq a$ (a presentation of this form exists in view of Remark 8.18). Define $q : \rho \to D$, in the following way. If $(a,b) \in \rho$ and $\varphi(a) = n(= \varphi(b))$, then a belongs to an R-class X_i and b belongs to a different R-class X_j of $Z(n)$. From the way ρ is constructed, $k = \min\{l \in \{1,\ldots,e\} \mid n_l \in A_j\} > 1$. Define $g((a,b)) = (n - n_k, k)$. As in the proof of Lemma 8.24, from the minimality of k, we can deduce that $w = n - n_k \in \mathrm{Ap}(S,n_1) \cap \cdots \cap \mathrm{Ap}(S,n_{k-1})$ and that it fulfills Conditions 1) and 2) of Lemma 8.24. From the way ρ is constructed, it follows that g is injective. Thus the cardinality of ρ is less than or equal to that of D. By Corollary 8.13, all minimal presentations have the same cardinality, and this concludes the proof. □

Thus the natural step to give is finding an upper bound for the cardinalities of the sets $D_1,\ldots,D_e$.

Theorem 8.26. *Let S be a numerical semigroup minimally generated by $\{n_1,\ldots,n_e\}$. The cardinality of any minimal presentation for S is less than or equal to*

$$\frac{(2n_1 - e + 1)(e - 2)}{2} + 1.$$

Proof. Define $c_e = \min\{k \in \mathbb{N} \setminus \{0\} \mid kn_e \in \langle n_1,\ldots,n_{e-1}\rangle\}$. Clearly $D_e = \{(c_e - 1)n_e\}$ and thus it has just one element. Besides,

$$D_k \subseteq (\mathrm{Ap}\,(S,n_1) \cap \cdots \cap \mathrm{Ap}\,(S,n_{k-1})) \setminus \{0\}.$$

As the cardinality of $\mathrm{Ap}\,(S,n_1)$ is n_1 (Lemma 2.4), and $n_2,\ldots,n_{k-1} \in \mathrm{Ap}\,(S,n_1)$ but they do not belong to $\mathrm{Ap}\,(S,n_1) \cap \cdots \cap \mathrm{Ap}\,(S,n_{k-1})$, we have that the cardinality of D_k is less than or equal to $n_1 - (k-1)$. The sum of the cardinalities of the sets $D_2,\ldots,D_e$ is then bounded by $1 + \sum_{k=2}^{e-1}(n_1 - k + 1) = 1 + \frac{(2n_1 - e + 1)(e-2)}{2}$. $\square$

By Proposition 2.10, e is less than or equal to n_1. We use this to give a weaker version of the above bound.

Corollary 8.27. *Let S be a numerical semigroup. The cardinality of any minimal presentation of S is less than or equal to*

$$\frac{\mathrm{m}(S)(\mathrm{m}(S) - 1)}{2}.$$

Proof. By Theorem 8.26, we know that the cardinality of any minimal presentation of S is less than or equal to $\frac{(2\mathrm{m}(S) - \mathrm{e}(S) + 1)(\mathrm{e}(S) - 2)}{2} + 1$. Observe that

$$\frac{(2\mathrm{m}(S) - \mathrm{e}(S) + 1)(\mathrm{e}(S) - 2)}{2} + 1 \leq \frac{\mathrm{m}(S)(\mathrm{m}(S) - 1)}{2}$$

if and only if $(\mathrm{m}(S) - \mathrm{e}(S))(\mathrm{m}(S) - \mathrm{e}(S) + 3) \geq 0$. This last equality holds since by Proposition 2.10, $\mathrm{e}(S) \leq \mathrm{m}(S)$. $\square$

Recall that a numerical semigroup S is said to have maximal embedding dimension if $\mathrm{e}(S) = \mathrm{m}(S)$. One should expect a simpler result for this class of numerical semigroups. In fact, as we will see next, numerical semigroups having maximal embedding dimension can be characterized as those for which the bound in Corollary 8.27 is reached, and thus with respect to the multiplicity they are those numerical semigroups with the largest possible presentations.

Lemma 8.28. *Let S be a numerical semigroup. If the cardinality of a minimal presentation of S is $\frac{\mathrm{m}(S)(\mathrm{m}(S)-1)}{2}$, then S has maximal embedding dimension.*

Proof. From the proof of Corollary 8.27 it easily follows that if the cardinality of a minimal presentation of S is $\frac{\mathrm{m}(S)(\mathrm{m}(S)-1)}{2}$, then $(\mathrm{m}(S) - \mathrm{e}(S))(\mathrm{m}(S) - \mathrm{e}(S) + 3) = 0$. Then either $\mathrm{e}(S) = \mathrm{m}(S)$ or $\mathrm{e}(S) = \mathrm{m}(S) + 3$. This last possibility cannot hold since $\mathrm{e}(S) \leq \mathrm{m}(S)$ in view of Proposition 2.10. $\square$

This proves one direction of the above-mentioned characterization. Let us proceed with the other implication.

Lemma 8.29. *Let S be a maximal embedding dimension numerical semigroup. Then the cardinality of any of its minimal presentations is $\frac{\mathrm{m}(S)(\mathrm{m}(S)-1)}{2}$.*

Proof. Set $m = \mathrm{m}(S)$. As S has maximal embedding dimension, S is minimally generated by $\{m, n_1, \ldots, n_{m-1}\}$ for some positive integers $n_1, \ldots, n_{m-1}$ and $\mathrm{Ap}\,(S,m) =$

$\{0, n_1, \ldots, n_{m-1}\}$ (see Proposition 3.1). By particularizing Proposition 8.19 to this setting, we have that if n is an element in S for which G_n is not connected, then $n = n_i + n_j$ for some $i, j \in \{1, \ldots, m-1\}$. We distinguish two cases.

1) If $i = j$, then $n = 2n_i$. Observe that $2n_i \notin \mathrm{Ap}(S, m)$, which in particular implies that $n - m \in S$. Hence n_i is the only vertex of a connected component, and m belongs to a different connected component.

2) If $i \neq j$, then $n = n_i + n_j$. Again $n_i + n_j \notin \mathrm{Ap}(S, m)$, and thus m is not in the same connected component containing n_i and n_j (a component which is just the edge $\overline{n_i n_j}$).

In view of 1) and 2), if G_n is not connected, then $m \in V_n$. Take n_k to be a vertex of G_n that is not in the same connected component of m. Then $n - n_k \in \mathrm{Ap}(S, m)$, that is, $n = n_k + n_l$ for some $l \in \{1, \ldots, m-1\}$. This proves that G_n is not connected if and only if $n = n_i + n_j$ for some $i, j \in \{1, \ldots, m-1\}$. Moreover, m is in a different connected component of n_i and n_j. Hence the cardinality of a minimal presentation coincides with the cardinality of the set $\{\{i, j\} \mid i, j \in \{1, \ldots, m-1\}\}$, which is equal to $\frac{m(m-1)}{2}$.   □

Theorem 8.30. *A numerical semigroup S has maximal embedding dimension if and only if the cardinality of any of its minimal presentations is* $\frac{m(S)(m(S)-1)}{2}$.

Exercises

Exercise 8.1 ([79]). Prove that the cardinality of a minimal presentation of $\langle 7, 8, 10, 19 \rangle$ is 7.

Exercise 8.2 ([4]). Show that $\langle 19, 23, 29, 31, 37 \rangle$ is symmetric and that any minimal presentation for it has cardinality 13.

Exercise 8.3 ([57]). Let S be a numerical semigroup minimally generated by $\{n_1 < \cdots < n_e\}$ with $\mathrm{F}(S) > \mathrm{m}(S)$. Let $T = S \cup \{\mathrm{F}(S)\}$. Assume that $\{n_1, \ldots, n_e, \mathrm{F}(S)\}$ is a minimal system of generators of T. Let $n \in S$. Denote by $\mathrm{ncc}(S, n)$ (respectively $\mathrm{ncc}(T, n)$) the number of connected components of the graph associated to n in S (respectively T). Prove that the following assertions are equivalent.

a) $\mathrm{ncc}(S, n) < \mathrm{ncc}(T, n)$.
b) $n \in \{\mathrm{F}(S) + n_1, \ldots, \mathrm{F}(S) + n_e, 2\mathrm{F}(S)\}$.
c) $\mathrm{ncc}(T, n) = \mathrm{ncc}(S, n) + 1$.

Prove also that the following are equivalent.

a) $\mathrm{ncc}(T, n) < \mathrm{ncc}(S, n)$.
b) $n = \mathrm{F}(S) + n_1 + n_i$ with $i \in \{2, \ldots, e\}$ and such that n_1 and n_i are in different connected components of the graph associated to n in S.
c) $\mathrm{ncc}(S, n) = \mathrm{ncc}(T, n) + 1$.

Exercise 8.4 ([57]). Let S be a numerical semigroup minimally generated by $\{n_1 < \cdots < n_e\}$ with $F(S) > m(S)$. Let $T = S \cup \{F(S)\}$. Assume that $\{n_1, \ldots, n_e, F(S)\}$ is not a minimal system of generator of T. By using the same notation as in Exercise 8.3, prove that the following assertions are equivalent.

a) $ncc(S,n) < ncc(T,n)$.
b) $n \in \{F(S) + n_2, \ldots, F(S) + n_{e-1}, 2F(S)\}$.
c) $ncc(T,n) = ncc(S,n) + 1$.

Prove also that the following are equivalent.

a) $ncc(T,n) < ncc(S,n)$.
b) $n \in \{n_e + n_2, \ldots, n_e + n_{e-1}, 2n_e\}$.
c) $ncc(S,n) = ncc(T,n) + 1$.

Exercise 8.5 ([57]). Let S be a numerical semigroup with $F(S) > m(S)$. Set $T = S \cup \{F(S)\}$. Assume that s and t are the cardinality of minimal presentations of S and T, respectively. Prove that

a) if $e(T) = e(S) + 1$, then $s + 2 \le t \le s + e(S) + 1$,
b) if $e(T) = e(S)$, then $s = t$.

Exercise 8.6 ([57]). Prove that if S is a numerical semigroup and ρ is a minimal presentation of S, then

$$\#\rho \le \frac{m(S)(m(S) - 1)}{2} - 2(m(S) - e(S)).$$

(*Hint:* Consider the set of semigroups in $Ch(S) = \{S_0 = S, S_1, \ldots, S_k = \mathbb{N}\}$. For some $m \in \{1, \ldots, k\}$, $S_m = \{0, m(S), \rightarrow\}$, which is a maximal embedding dimension numerical semigroup. Then use the preceding exercise.)

Exercise 8.7 ([79]). Let S be a numerical semigroup with $e(S) = m(S) - 1$ and let ρ be a minimal presentation for S. Show that

$$\frac{(m(S) - 1)(m(S) - 2)}{2} - 1 \le \#\rho \le \frac{(m(S) - 1)(m(S) - 2)}{2}.$$

Exercise 8.8 ([79]). Let S be a symmetric numerical semigroup with $e(S) \in \{m(S) - 1, m(S) - 2, m(S) - 3\}$ and let ρ be a minimal presentation for S. Prove that

$$\#\rho = \frac{e(S)(e(S) - 1)}{2} - 1.$$

Exercise 8.9 ([73]). Let S be an irreducible numerical semigroup with embedding dimension greater than four. Prove that the cardinality of any minimal presentation of S is less than or equal to

$$\frac{(m(S) - 2)(m(S) - 1)}{2} - (m(S) - e(S)).$$

Exercise 8.10 ([33]). Let a and b be integers with $1 \leq b < a$. Assume that ρ is a minimal presentation of $\langle a, a+1, \ldots, a+b \rangle$. Prove that

$$\#\rho = \frac{b(b-1)}{2} + b - ((a-1) \bmod b).$$

Exercise 8.11. Let σ be the congruence on $\mathbb{N}$ generated by $\{(1,2)\}$. Prove that σ is not finitely generated as a submonoid of $\mathbb{N} \times \mathbb{N}$.

Exercise 8.12. Let S be a numerical semigroup minimally generated by $\{n_1, \ldots, n_e\}$. A *factorization* of $n \in S$ is an element $a = a_1 x_1 + \cdots + a_e x_e$ of $\mathsf{Z}(n)$. The *length* of a is $|a| = a_1 + \cdots + a_n$. A finitely generated monoid is *half-factorial* if for every element in the monoid, all its factorizations have the same length. Prove that the only half-factorial numerical semigroup is $\mathbb{N}$.

Exercise 8.13 ([20]). Let S be a numerical semigroup minimally generated by $\{n_1, \ldots, n_e\}$ and let $n \in S$. Let $a, b \in \mathsf{Z}(n)$, with $a = a_1 x_1 + \cdots + a_e x_e$ and $b = b_1 x_1 + \cdots + b_e x_e$. The *greatest common divisor* of a and b is defined as

$$\gcd\{a, b\} = (\min\{a_1, b_1\}) x_1 + \cdots + (\min\{a_e, b_e\}) x_e.$$

The *distance* between a and b is

$$\mathrm{dist}(a, b) = \max\{|a - \gcd\{a, b\}|, |b - \gcd\{a, b\}|\}$$

(see [36, Proposition 1.2.5] for a list of basic properties concerning the distance).

Given a positive integer N, an N-chain of factorizations from a to b is a sequence $z_0, \ldots, z_k \in \mathsf{Z}(n)$ such that $z_0 = a$, $z_k = b$ and $\mathrm{dist}(z_i, z_{i+1}) \leq N$ for all i. The *catenary degree* of n, $\mathrm{cd}(n)$, is the minimal $N \in \mathbb{N} \cup \{\infty\}$ such that for any two factorizations $a, b \in \mathsf{Z}(n)$, there is an N-chain from a to b. Let ρ be a minimal presentation. Prove that $\mathrm{cd}(n) \leq \max\{\max\{|a|, |b|\} \mid (a, b) \in \rho\}$.

Chapter 8
The gluing of numerical semigroups

Introduction

If $K[V]$ is the ring of coordinates of a variety V, then V is said to be a complete intersection if its defining ideal is generated by the least possible number of polynomials. In the special case $K[V]$ is taken to be a semigroup ring $K[S]$, the generators of its defining ideal can be chosen to be binomials whose exponents correspond to a presentation of the monoid S (see [41]). In this way the concept of complete intersection translates to finitely generated monoids as those having the least possible number of relations in their minimal presentations.

The concept of gluing of numerical semigroups was introduced in Rosales' PhD thesis ([55]) in order to prove that a numerical semigroup is a complete intersection if and only if it is the gluing of two complete intersection numerical semigroups, rewriting in this way Delorme's characterization of complete intersection numerical semigroups ([27]). Later the definition of gluing is generalized for affine semigroups (finitely generated submonoids of $\mathbb{N}^n$) in [60]. These results were then used in [77] to generalize the characterization of complete intersection numerical semigroups to submonoids of $\mathbb{N}^2$, $\mathbb{N}^3$ and any simplicial affine semigroup. Finally, in [30] the idea of gluing is used to prove that an affine semigroup is a complete intersection if and only if it is the gluing of two complete intersection affine semigroups.

Complete intersection numerical semigroups are symmetric, and thus they provide examples of one-dimensional Gorenstein local domains via their semigroup rings. This is why some authors have given procedures to construct complete intersection numerical semigroups. Bertin and Carbonne in [6] introduce a subclass of complete intersection numerical semigroups which they call free semigroups (this has nothing to do with the concept of categorical freeness, since the only free numerical semigroup in this sense is $\mathbb{N}$). These semigroups have later been used by other authors to produce examples of one-dimensional Gorenstein local domains with a given multiplicity and embedding dimension (see [107]), and also algebraic codes (see [43]).

J.C. Rosales, P.A. García-Sánchez, *Numerical Semigroups*,
Developments in Mathematics 20, DOI 10.1007/978-1-4419-0160-6_9,
© Springer Science+Business Media, LLC 2009

1 The concept of gluing

The idea of gluing is the following. A set of positive integers A, which is usually taken as the set of generators of a monoid, is the gluing of A_1 and A_2 if $\{A_1, A_2\}$ is a partition of A and the monoid generated by A admits a presentation in which some relators only involve generators in A_1, other relators only involve generators in A_2 and there is only one element in this presentation relating elements in A_1 with elements in A_2. In order to formalize this definition we need to recall and introduce some notation.

Let $A = \{m_1, \ldots, m_r\}$ be a subset of positive integers. Let $X = \{x_1, \ldots, x_r\}$. Let $\varphi : \mathrm{Free}(X) \to \mathbb{N}$ the monoid homomorphism

$$\varphi(a_1 x_1 + \ldots + a_r x_r) = a_1 m_1 + \cdots + a_r m_r.$$

Denote by σ the kernel congruence of φ, that is, $a \sigma b$ if and only if $\varphi(a) = \varphi(b)$. For $B \subseteq A$, set $X_B = \{x_i \mid m_i \in B\}$. Then $\mathrm{Free}(X_B) \subseteq \mathrm{Free}(X)$. We define φ_B and σ_B accordingly. Note that $\sigma_B \subseteq \sigma$.

With this notation it is now easy to express the concept of gluing. Let $\{A_1, A_2\}$ be a partition of A. We say that A is the *gluing* of A_1 and A_2 if there exists a system of generators ρ of σ such that $\rho = \rho_1 \cup \rho_2 \cup \{(a,b)\}$ with $\rho_1 \subseteq \sigma_{A_1}$, $\rho_2 \subseteq \sigma_{A_2}$, $0 \neq a \in \mathrm{Free}(X_{A_1})$ and $0 \neq b \in \mathrm{Free}(X_{A_2})$.

The Apéry set of an element with respect to a set of integers can be defined in the following way. If C is a subset of $\mathbb{N}$ and $x \in \langle C \rangle \setminus \{0\}$ denote by

$$\mathrm{Ap}(C,x) = \{x \in \langle C \rangle \mid s - x \notin \langle C \rangle\}.$$

Clearly, if $y \in \langle C \rangle$, then there exists a unique $(s,k) \in \mathrm{Ap}(C,x) \times \mathbb{N}$ such that $y = s + kx$ (compare with Lemma 2.6).

The following technical lemma will be helpful to control $\langle A_1 \rangle \cap \langle A_2 \rangle$ in order to give a characterization of gluing.

Lemma 9.1. *Let $\rho = \rho_1 \cup \rho_2 \cup \{(a,b)\}$ be a system of generators of σ, with $\rho_1 \subseteq \sigma_{A_1}$, $\rho_2 \subseteq \sigma_{A_2}$, $0 \neq a \in \mathrm{Free}(X_{A_1})$ and $0 \neq b \in \mathrm{Free}(X_{A_2})$. Let $s_1, t_1 \in \mathrm{Ap}(A_1, \varphi(a))$, $s_2, t_2 \in \mathrm{Ap}(A_2, \varphi(a))$ and $k \in \mathbb{N}$ such that $s_1 + s_2 = k\varphi(a) + t_1 + t_2$. Then $k = 0$, $s_1 = t_1$ and $s_2 = t_2$.*

Proof. Assume for the sake of simplicity and without loss of generality that $A_1 = \{m_1, \ldots, m_h\}$ and $A_2 = \{m_{h+1}, \ldots, m_r\}$. Observe that $0 \neq \varphi(a) = \varphi(b) \in \langle A_1 \rangle \cap \langle A_2 \rangle$. Thus, $\varphi(a) = a_1 m_1 + \cdots + a_h m_h = a_{h+1} m_{h+1} + \cdots + a_r m_r$ for some $a_1, \ldots, a_r \in \mathbb{N}$. As $s_1, t_1 \in \langle A_1 \rangle$ and $s_2, t_2 \in \langle A_2 \rangle$, there exist $d_1, \ldots, d_r, b_1, \ldots, b_r \in \mathbb{N}$ such that $s_1 = d_1 m_1 + \cdots + d_h m_h$, $t_1 = b_1 m_1 + \cdots + b_h m_h$, $s_2 = d_{h+1} m_{h+1} + \cdots + d_r m_r$ and $t_2 = b_{h+1} m_{h+1} + \cdots + b_r m_r$.

By hypothesis $s_1 + s_2 = k\varphi(a) + t_1 + t_2$. Then

$$(d_1 x_1 + \cdots + d_r x_r) \sigma ((ka_1 + b_1) x_1 + \cdots + (ka_h + b_h) x_h + b_{h+1} x_{h+1} + \cdots + b_r x_r).$$

Since ρ is a system of generators of σ, by Proposition 8.4, there exist $v_0, \ldots, v_t \in$ Free(X) such that

- $v_0 = d_1 x_1 + \cdots + d_r x_r$,
- $v_t = (ka_1 + b_1)x_1 + \cdots + (ka_h + b_h)x_h + b_{h+1}x_{h+1} + \cdots + b_r x_r$ and
- $(v_i, v_{i+1}) \in \rho^1$ for all $i \in \{0, \ldots, t-1\}$.

We prove by induction on i that if $v_i = v_{i_1}x_1 + \cdots + v_{i_r}x_r$, for some $v_{i_1}, \ldots, v_{i_r} \in \mathbb{N}$, then $s_1 = v_{i_1}m_1 + \cdots + v_{i_h}m_h$ and $s_2 = v_{i_{h+1}}m_{h+1} + \cdots + v_r m_{i_r}$.

The result is clear for $i = 0$. Assume that the result holds for i and let us prove it for $i+1$. As $(v_i, v_{i+1}) \in \rho^1$, there exist $(u, v) \in \rho \cup \rho^{-1} \cup \Delta(\text{Free}(X))$ and $w = w_1 x_1 + \cdots + w_r x_r \in$ Free(X) such that $(v_i, v_{i+1}) = (u+w, v+w)$ (see Proposition 8.4). We distinguish five cases.

1) If $(u,v) \in \Delta(\text{Free}(X))$, then the result is true, because in this setting $v_i = v_{i+1}$.
2) If $(u,v) = (a,b)$, then as $v_i = u+w = a+w$, $s_1 = v_{i_1}m_1 + \cdots + v_{i_h}m_h = \varphi(a) + w_1 m_1 + \cdots + w_r m_r$. However this implies that $s_1 \notin \text{Ap}(A_1, \varphi(a))$, contradicting one of the hypotheses. Thus this case cannot occur.
3) The same stands for $(u,v) = (b,a)$.
4) If $(u,v) \in \rho_1 \cup \rho_1^{-1}$, then there exist $c_1, \ldots, c_h, e_1, \ldots, e_h \in \mathbb{N}$ such that $(u,v) = (c_1 x_1 + \cdots + c_h x_h, e_1 x_1 + \cdots + e_h x_h)$. Hence

$$v_i = (c_1 + w_1)x_1 + \cdots + (c_h + w_h)x_h + w_{h+1}x_{h+1} + \cdots + w_r x_r$$

and

$$v_{i+1} = (e_1 + w_1)x_1 + \cdots + (e_h + w_h)x_h + w_{h+1}x_{h+1} + \cdots + w_r x_r.$$

As $(u,v) \in \sigma$, we have that $c_1 m_1 + \cdots + c_r m_r = e_1 m_1 + \cdots + e_r m_r$. Thus $v_{i+1_1}m_1 + \cdots + v_{i+1_h}m_h = v_{i_1}m_1 + \cdots + v_{i_h}m_h = s_1$ and $v_{i+1_{h+1}}m_{h+1} + \cdots + v_{i+1_r}m_r = v_{i_{h+1}}m_{h+1} + \cdots + v_{i_r}m_r = s_2$.
5) The case $(u,v) \in \rho_2 \cup \rho_2^{-1}$ is analogous to the preceding case.

As $v_t = (ka_1 + b_1)x_1 + \cdots + (ka_h + b_h)x_h + b_{h+1}x_{h+1} + \cdots + b_r x_r$, we obtain that $s_1 = (ka_1 + b_1)m_1 + \cdots + (ka_h + b_h)m_h$ and $s_2 = b_{h+1}m_{h+1} + \cdots + b_r m_r$. It is clear that $k = 0$ because $s_1 \in \text{Ap}(A_1, \varphi(a))$ and $\varphi(a) = a_1 m_1 + \cdots + a_h m_h$. Hence $k = 0$, $s_1 = t_1$ and $s_2 = t_2$. $\qquad\square$

The concept of gluing can be characterized in several ways. We collect here some of the given in [55]. These will enable us to define the gluing of two numerical semigroups and control the notable elements as well as the presentation of the gluing from the original semigroups.

Theorem 9.2. *Let A be a subset of positive integers. Let $\{A_1, A_2\}$ be a partition of A. The following assertions are equivalent.*

1) A is the gluing of A_1 and A_2.
2) If $d_1 = \gcd(A_1)$ and $d_2 = \gcd(A_2)$, then $\text{lcm}\{d_1, d_2\} \in \langle A_1 \rangle \cap \langle A_2 \rangle$ (where lcm stands for least common multiple).

3) There exists $d \in \langle A_1 \rangle \cap \langle A_2 \rangle$, $d \neq 0$, such that the correspondence

$$f : \mathrm{Ap}(A_1, d) \times \mathrm{Ap}(A_2, d) \to \mathrm{Ap}(A, d), \quad f(s_1, s_2) = s_1 + s_2$$

is a bijective map.
4) There exists $(a, b) \in \sigma$ with $0 \neq a \in \mathrm{Free}(X_{A_1})$ and $0 \neq b \in \mathrm{Free}(X_{A_2})$ such that $\rho_1 \cup \rho_2 \cup \{(a, b)\}$ is a system of generators of σ, for every ρ_1 and ρ_2 systems of generators of σ_{A_1} and σ_{A_2}, respectively.

Proof. 1) implies 2). Let $d = \varphi(a)$. As $(a, b) \in \sigma$, $0 \neq a \in \mathrm{Free}(X_{A_1})$ and $0 \neq b \in \mathrm{Free}(X_{A_2})$, we deduce that $d \in \langle A_1 \rangle \cap \langle A_2 \rangle$ and $d \neq 0$. We prove that $d = \mathrm{lcm}\{d_1, d_2\}$. Since $d \in \langle A_1 \rangle \cap \langle A_2 \rangle$, d is a multiple of d_1 and d_2. Hence d is a multiple of $\mathrm{lcm}\{d_1, d_2\}$. By using Bézout's identity, we can easily derive that there exist $y_i, z_i \in \langle A_i \rangle$ for $i \in \{1, 2\}$ such that $\mathrm{lcm}\{d_1, d_2\} = y_1 - z_1 = y_2 - z_2$. Hence there exist $k_1, l_1, k_2, l_2 \in \mathbb{N}$, $s_1, t_1 \in \mathrm{Ap}(A_1, d)$ and $s_2, t_2 \in \mathrm{Ap}(A_2, d)$ such that $y_i = k_i d + s_i$ and $z_i = l_i d + t_i$ for $i \in \{1, 2\}$. By using that $y_1 - z_1 = y_2 - z_2$, we obtain that $(k_1 + l_2)d + s_1 + t_2 = (k_2 + l_1)d + t_1 + s_2$. Assume without loss of generality that $k_1 + l_2 \leq k_2 + l_1$. Then $s_1 + t_2 = (k_2 + l_1 - k_1 - l_2)d + t_1 + s_2$. By Lemma 9.1, we have that $s_1 = t_1$ and that $s_2 = t_2$. Hence $\mathrm{lcm}\{d_1, d_2\} = y_1 - z_1 = k_1 d + s_1 - l_1 d - t_1 = (k_1 - l_1)d$. This proves that $\mathrm{lcm}\{d_1, d_2\}$ is a multiple of d and consequently $d = \mathrm{lcm}\{d_1, d_2\}$.

2) implies 3). Let $d = \mathrm{lcm}\{d_1, d_2\}$. We first prove that if $(s_1, s_2) \in \mathrm{Ap}(A_1, d) \times \mathrm{Ap}(A_2, d)$, then $s_1 + s_2 \in \mathrm{Ap}(A, d)$. As $s_1 + s_2 \in \langle A \rangle$, there exists $(s, k) \in \mathrm{Ap}(A, d) \times \mathbb{N}$ such that $s_1 + s_2 = kd + s$. Since $s \in \langle A \rangle$, there exists nonnegative integers $b_1, \ldots, b_r$ such that $s = b_1 m_1 + \ldots + b_r m_r$. It follows that $t_1 = b_1 m_1 + \cdots + b_h m_h \in \mathrm{Ap}(A_1, d)$ and that $t_2 = b_{h+1} m_{h+1} + \cdots + b_r m_r \in \mathrm{Ap}(A_2, d)$. Then $s_1 + s_2 = kd + t_1 + t_2$. Hence $s_1 - t_1 = kd + s_2 - t_2$. Observe that $s_1 - t_1$ is a multiple of d_1 and $kd + s_2 - t_2$ is a multiple of d_2. Consequently, $s_1 - t_1 = kd + s_2 - t_2 = zd$ for some integer z. As both s_1 and t_1 belong to $\mathrm{Ap}(A_1, d)$, this equality forces z to be zero. Thus $kd + s_2 = t_2$, and this again forces k to be zero, since otherwise $t_2 \notin \mathrm{Ap}(A_2, d)$. This proves that $s = s_1 + s_2 \in \mathrm{Ap}(A, d)$.

Now let us show that f is injective. Assume that $f(s_1, s_2) = f(t_1, t_2)$. Then $s_1 - t_1 = s_2 - t_2$. Arguing as above, there exists an integer z such that $s_1 - t_1 = s_2 - t_2 = zd$. And again this forces z to be zero, which leads to $s_1 = t_1$ and $s_2 = t_2$.

Finally take $s = b_1 m_1 + \cdots + b_r m_r \in \mathrm{Ap}(A, d)$. Then setting $s_1 = b_1 m_1 + \cdots + b_h m_h$ and $s_2 = b_{h+1} m_{h+1} + \cdots + b_r m_r$, we have that $(s_1, s_2) \in \mathrm{Ap}(A_1, d) \times \mathrm{Ap}(A_2, d)$ and $f(s_1, s_2) = s$.

3) implies 4). As $d \in \langle A_1 \rangle \cap \langle A_2 \rangle$, there exist $a_1, \ldots, a_r \in \mathbb{N}$ such that $d = a_1 m_1 + \cdots + a_h m_h = a_{h+1} m_{h+1} + \cdots + a_r m_r$. Let ρ_1 and ρ_2 be systems of generators of σ_{A_1} and σ_{A_2}, respectively. Let $a = a_1 x_1 + \cdots + a_h x_h$ and $b = a_{h+1} x_{h+1} + \cdots + a_r x_r$. Then $0 \neq a \in \mathrm{Free}(X_{A_1})$, $0 \neq b \in \mathrm{Free}(X_{A_2})$ and $\varphi(a) = \varphi(b) = d$. Hence $(a, b) \in \sigma$. We prove that $\rho = \rho_1 \cup \rho_2 \cup \{(a, b)\}$ is a system of generators of σ. As $\rho \subseteq \sigma$, $\mathrm{Cong}(\rho) \subseteq \sigma$. Take $(b_1 x_1 + \cdots + b_r x_r, c_1 x_1 + \cdots + c_r x_r) \in \sigma$. Since $b_1 m_1 + \cdots + b_h m_h \in \langle A_1 \rangle$, there exists $(s_1, k_1) \in \mathrm{Ap}(A_1, d) \times \mathbb{N}$ such that $b_1 m_1 + \cdots + b_h m_h = k_1 d + s_1$. Arguing analogously, there exist $(t_1, l_1) \in \mathrm{Ap}(A_1, d)$ and $(s_2, k_2), (t_2, l_2) \in \mathrm{Ap}(A_2, d)$ such that $b_{h+1} m_{h+1} + \cdots + b_r m_r = k_2 d + s_2$, $c_1 m_1 + \cdots + c_h m_h = l_1 d +$

t_1 and $c_{h+1}m_{h+1} + \cdots + c_r m_r = l_2 d + t_2$. From $b_1 m_1 + \cdots + b_r m_r = c_1 m_1 + \cdots + c_r m_r$, it follows that $(k_1 + k_2)d + s_1 + s_2 = (l_1 + l_2)d + t_1 + t_2$. We deduce that $k_1 + k_2 = l_1 + l_2$, because $s_1 + s_2, t_1 + t_2 \in \text{Ap}(S,d)$. Hence $s_1 + s_2 = t_1 + t_2$. By using now that f is injective, $s_1 = t_1$ and $s_2 = t_2$. Let $g_1, \ldots, g_r \in \mathbb{N}$ be such that $s_1 = t_1 = g_1 m_1 + \cdots + g_h m_h$ and $s_2 = t_2 = g_{h+1} m_{h+1} + \cdots + g_r m_r$. Then $(b_1 x_1 + \cdots + b_h x_h, (k_1 a_1 + g_1)x_1 + \cdots + (k_1 a_h + g_h)x_h) \in \sigma_{A_1}$ (because $b_1 m_1 + \cdots + b_h m_h = k_1 d + s_1$), and thus $(b_1 x_1 + \cdots + b_h x_h, (k_1 a_1 + g_1)x_1 + \cdots + (k_1 a_h + g_h)x_h) \in \text{Cong}(\rho_1) \subseteq \text{Cong}(\rho)$. Analogously, $(b_{h+1}x_{h+1} + \cdots + b_r x_r, (k_2 a_{h+1} + g_{h+1})x_{h+1} + \cdots + (k_2 a_r + g_r)x_r) \in \text{Cong}(\rho_2) \subseteq \text{Cong}(\rho)$. Taking into account that $\text{Cong}(\rho)$ is a congruence, $(b_1 x_1 + \cdots + b_r x_r, (k_1 a_1 + g_1)x_1 + \cdots + (k_1 a_h + g_h)x_h + (k_2 a_{h+1} + g_{h+1})x_{h+1} + \cdots + (k_2 a_r + g_r)x_r) \in \text{Cong}(\rho)$. In the same way, we obtain that $(c_1 x_1 + \cdots + c_r x_r, (l_1 a_1 + g_1)x_1 + \cdots + (l_1 a_h + g_h)x_h + (l_2 a_{h+1} + g_{h+1})x_{h+1} + \cdots + (l_2 a_r + g_r)x_r) \in \text{Cong}(\rho)$.

Let $\zeta = \text{Cong}(\{(a,b)\})$. As $k_1 + k_2 = l_1 + l_2$, $(k_1 a + k_2 b)\zeta(k_1 + k_2)b\zeta(l_1 + l_2)b\zeta l_1 a + l_2 b$. Hence $(k_1 a + k_2 b, l_1 a + l_2 b) \in \text{Cong}(\{(a,b)\}) \subseteq \text{Cong}(\rho)$. We can then conclude that $(b_1 x_1 + \cdots + b_r x_r, c_1 x_1 + \cdots + c_r x_r) \in \text{Cong}(\rho)$.

4) implies 1). Trivial.   □

This theorem can be used sometimes to compute presentations of numerical semigroups as the following example shows.

Example 9.3. Let S be the numerical semigroup (minimally) generated by $A = \{85, 187, 221, 60, 80, 90\}$. Let $A_1 = \{85, 187, 221\}$ and $A_2 = \{60, 80, 90\}$. Then $\gcd(A_1) = 17$, $\gcd(A_2) = 10$ and $\text{lcm}\{10, 17\} = 170 \in \langle A_1 \rangle \cap \langle A_2 \rangle$. Hence A is the gluing of A_1 and A_2. The monoid $\langle A_1 \rangle$ is isomorphic to $\langle A_1/17 \rangle = \langle 5, 11, 13 \rangle$. A system of generators for σ_{A_1} is $\{(7x_1, 2x_2 + x_3), (3x_2, 4x_1 + x_3), (2x_3, 3x_1 + x_2)\}$ (see Example 8.23). Furthermore, $\langle A_2 \rangle \cong \langle A_2/10 \rangle = \langle 6, 8, 9 \rangle$. A minimal presentation for σ_{A_2} is $\{(4x_4, 3x_5), (3x_4, 2x_6)\}$ (again by Example 8.23). Moreover, $170 = 2 \times 85 = 80 + 90$, whence we can choose $(a,b) = (2x_1, x_5 + x_6)$ and obtain by the preceding theorem that

$$\rho = \{(7x_1, 2x_2 + x_3), (3x_2, 4x_1 + x_3), (2x_3, 3x_1 + x_2),$$
$$(4x_4, 3x_5), (3x_4, 2x_6), (2x_1, x_5 + x_6)\}$$

is a presentation for S.

2 Complete intersection numerical semigroups

In this section we give a lower bound for the cardinality of a presentation of a numerical semigroup. Then we will characterize those numerical semigroups for which this bound is reached. Recall that we already know an upper bound for the cardinality of a minimal presentation of a numerical semigroup, and that this bound was reached for (and only for) numerical semigroups with maximal embedding dimension (see Section 4 of Chapter 7).

We assume that S is a numerical semigroup minimally generated by $A = \{n_1, \ldots, n_e\}$ and that $X = \{x_1, \ldots, x_e\}$. As in the preceding section $\varphi : \text{Free}(X) \to \mathbb{N}$ denotes the monoid morphism defined as $\varphi(a_1 x_1 + \cdots + a_e x_e) = a_1 n_1 + \cdots + a_e n_e$, and σ is its kernel congruence. Let ρ be a system of generators of σ.

Given B a nonempty proper subset of A, denote by

$$\mu(B) = \min \{ x \in \langle B \rangle \mid x - a \in S \text{ for some } a \in A \setminus B \}.$$

This minimum always exist.

For $s \in S$, recall that $X_B = \{ x_i \in X \mid n_i \in B \}$. Define

$$Z_B(s) = Z(s) \cap \text{Free}(X_B).$$

The following lemma ensures the existence of certain elements in a minimal presentation of any numerical semigroup.

Lemma 9.4. *Let B be a nonempty proper subset of A. Then there exist $e_B(\mu(B)) \in Z_B(\mu(B))$ and $e(\mu(B)) \in Z(\mu(B)) \setminus Z_B(\mu(B))$ such that $(e_B(\mu(B)), e(\mu(B))) \in \rho \cup \rho^{-1}$.*

Proof. Assume for the sake of simplicity (and without loss of generality) that $B = \{n_1, \ldots, n_r\}$ and that $\mu(B) - n_{r+1} \in S$. There exist $a_1, \ldots, a_r, b_1, \ldots, b_e \in \mathbb{N}$ such that $\mu(B) = a_1 n_1 + \cdots + a_r n_r = b_1 n_1 + \cdots + b_e n_e$ with $b_{r+1} \neq 0$. Write $a = a_1 x_1 + \cdots + a_r x_r$ and $b = b_1 x_1 + \cdots + b_e x_e$.

Let $d = d_1 x_1 + \cdots + d_r x_r \in Z_B(\mu(B))$. Let us see that if $c = c_1 x_1 + \cdots + c_e x_e \in Z(\mu(B))$ is such that $d \cdot c \neq 0$, then $c_{r+1} = \cdots = c_e = 0$. Assume to the contrary that $c_k \neq 0$ for some k greater than r. As $d \cdot c \neq 0$, there exist $i \in \{1, \ldots, r\}$ such that $c_i \neq 0 \neq d_i$. This implies $\mu(B) - n_i \in \langle B \rangle$, and also that $\mu(B) - (n_i + n_k) \in S$, contradicting the minimality of $\mu(B)$. It easily follows that

- if $c_1 x_1 + \ldots + c_e x_e$ is in the R-class of a, then $c_{r+1} = \cdots = c_e = 0$;
- thus, the elements a and b are in different R-classes of $Z(\mu(B))$ (because $b_{r+1} \neq 0$).

By using that ρ is a minimal system of generators of σ, the proof is a consequence of Lemma 8.9. $\qquad\square$

Recurrently define the pair (P_m, γ_m), with P_m a partition of A and γ_m a subset of ρ, as follows

- $P_1 = \{\{n_1\}, \ldots, \{n_e\}\}$ and $\gamma_1 = \emptyset$.
- once defined $P_m = \{B_1, \ldots, B_t\}$ and γ_m,
  - if there exists $(e_{B_i}(n), e_{B_j}(n)) \in \rho$ with $n \in \mathbb{N}$ and $i, j \in \{1, \ldots, t\}$, $i \neq j$, then set
  $$P_{m+1} = (P_m \setminus \{B_i, B_j\}) \cup \{B_i \cup B_j\},$$
  $$\gamma_{m+1} = \gamma_m \cup \{(e_{B_i}(n), e_{B_j}(n))\},$$

    – otherwise, set $P_{m+1} = P_m$ and $\gamma_{m+1} = \gamma_m$.

Let us study some of the properties of these pairs. They will be used to find the lower bound for the cardinality of a minimal presentation of a numerical semigroup, and later to understand the structure of those numerical semigroups reaching this bound.

Lemma 9.5. *Under the standing hypothesis and notation.*

1) $\#\gamma_m = e - \#P_m$,

2) if $P_m = \{B_1, \ldots, B_t\}$, then $\gamma_m \subseteq \sigma_{B_1} \cup \cdots \cup \sigma_{B_t}$,

3) if $\#\rho \leq e - 1$ and $\#P_m \geq 2$, then $\#P_{m+1} = \#P_m - 1$.

Proof. 1) and 2) follow directly from the definition. As for 3), if $P_m = \{B_1, \ldots, B_t\}$, by Lemma 9.4, we know that there exists $\xi = \{(e_{B_1}(\mu(B_1)), e(\mu(B_1))), \ldots, (e_{B_t}(\mu(B_t)), e(\mu(B_t)))\} \subseteq \rho \cup \rho^{-1}$. Let β be the set obtained by replacing (a, b) with (b, a) in ρ whenever $(a, b) \notin \rho$. Then $\gamma_m \cup \beta \subseteq \rho$. As $\gamma_m \subseteq \sigma_{B_1} \cup \cdots \cup \sigma_{B_t}$ and $\beta \cap (\sigma_{B_1} \cup \cdots \cup \sigma_{B_t})$ is empty, we have that $\#\gamma_m + \#\beta \leq \#\rho$. We are assuming that $\#\rho \leq e - 1$, which leads to $e - t + \#\beta \leq e - 1$, and consequently $\#\beta \leq t - 1$. We deduce then that there exist $i, j \in \{1, \ldots, t\}$, $i \neq j$ such that $(e_{B_i}(\mu(B_i)), e_{B_j}(\mu(B_j))) \in \beta \subseteq \rho$. Hence $\#P_{m+1} \leq \#P_m - 1$. $\square$

The lower bound can now be established.

Theorem 9.6. *Let S be a numerical semigroup. Then the cardinality of a minimal presentation for S is greater than or equal to $e(S) - 1$.*

Proof. Let $\{n_1, \ldots, n_e\}$ be a minimal generating system of S and let ρ be a minimal presentation for S. Assume to the contrary that $\#\rho < e - 1$. From Lemma 9.5 we deduce that $\#P_e = 1$ and that $\#\gamma_e = e - 1$. As $\gamma_e \subseteq \rho$, this forces $e - 1 \leq \#\rho$, a contradiction. $\square$

A numerical semigroup is a *complete intersection* if the cardinality of any of its minimal presentations equals its embedding dimension minus one.

Example 9.7. Observe that $\mathbb{N}$ is a complete intersection. Its multiplicity minus one is zero, and this is precisely the cardinality of any minimal presentation for $\mathbb{N}$.

If n_1 and n_2 are integers greater than one such that $\gcd\{n_1, n_2\} = 1$, then a minimal presentation for $\langle n_1, n_2 \rangle$ is $\{(n_2 x_1, n_1 x_2)\}$ (see Example 8.22). Thus $\langle n_1, n_2 \rangle$ is a complete intersection.

From Example 8.23, we deduce that complete intersection numerical semigroups of embedding dimension three $\langle n_1, n_2, n_3 \rangle$ are those fulfilling that $\#\{c_1 n_1, c_2 n_2, c_3 n_3\} \leq 2$.

3 Gluing of numerical semigroups

We translate the concept of gluing to numerical semigroups and prove that complete intersection numerical semigroups are always a gluing of two complete intersection numerical semigroups. As a consequence of this, we will see that every complete intersection numerical semigroup is symmetric.

Let S_1 and S_2 be two numerical semigroups minimally generated by $\{n_1, \ldots, n_r\}$ and $\{n_{r+1}, \ldots, n_e\}$, respectively. Let $\lambda \in S_1 \setminus \{n_1, \ldots, n_r\}$ and $\mu \in S_2 \setminus \{n_{r+1}, \ldots, n_e\}$ be such that $\gcd\{\lambda, \mu\} = 1$. We say that

$$S = \langle \mu n_1, \ldots, \mu n_r, \lambda n_{r+1}, \ldots, \lambda n_e \rangle$$

is a *gluing* of S_1 and S_2. We now see the connection between this concept and that of gluing of subsets of nonnegative integers.

Lemma 9.8. *Under the standing hypothesis,*

1) S is a numerical semigroup with minimal system of generators
   $\{\mu n_1, \ldots, \mu n_r, \lambda n_{r+1}, \ldots, \lambda n_e\}$,
2) $\{\mu n_1, \ldots, \mu n_r, \lambda n_{r+1}, \ldots, \lambda n_e\}$ is the gluing of $\{\mu n_1, \ldots, \mu n_r\}$ and
   $\{\lambda n_{r+1}, \ldots, \lambda n_e\}$.

Proof.
1) It is clear that $\gcd\{\mu n_1, \ldots, \mu n_r, \lambda n_{r+1}, \ldots, \lambda n_e\} = \gcd\{\mu, \lambda\} = 1$ and thus S is a numerical semigroup. Assume that $\mu n_1 = a_2 \mu n_2 + \cdots + a_r \mu n_r + a_{r+1} \lambda n_{r+1} + \cdots + a_e \lambda n_e$. Then $a_{r+1} \lambda n_{r+1} + \cdots + a_e \lambda n_e$ is a multiple of $\lambda \mu$. Hence $\mu n_1 = a_2 \mu n_2 + \cdots + a_r \mu n_r + \lambda \mu k$ for some nonnegative integer k. This leads to $n_1 = a_2 n_2 + \cdots + a_r n_r + \lambda k$. But this is impossible because $\lambda \in S_1$ and n_1 is a minimal generator of S_1. In the same way we prove that the rest of the elements in $\{\mu n_1, \ldots, \mu n_r, \lambda n_{r+1}, \ldots, \lambda n_e\}$ are minimal generators of S.
2) In view of Theorem 9.2, it suffices to prove that

$$\lambda \mu \in \langle \mu n_1, \ldots, \mu n_r \rangle \cap \langle \lambda n_{r+1}, \ldots, \lambda n_e \rangle.$$

This follows easily from the choice of λ and μ. $\square$

Now that we have the connection with these two concepts, with the help of Theorem 9.2 it is not hard to prove that a gluing of complete intersections yields a complete intersection numerical semigroup.

Proposition 9.9. *A gluing of two complete intersections is a complete intersection.*

Proof. Let S_1 and S_2 be complete intersection numerical semigroups, and let S be a gluing of S_1 and S_2. By Lemma 9.8, we know that $e(S) = e(S_1) + e(S_2)$. Let ρ_1 and ρ_2 be minimal presentations for S_1 and S_2, respectively. Then $\#\rho_i = e(S_i) - 1$ for $i \in \{1, 2\}$. By using Lemma 9.8 and Theorem 9.2, we obtain that S admits a presentation of the form $\rho = \rho_1 \cup \rho_2 \cup \{(a, b)\}$. Hence $\#\rho = \#\rho_1 + \#\rho_2 + 1 = e(S_1) - 1 + e(S_2) - 1 + 1 = e(S) - 1$. By using now Theorem 9.6 and Corollary 8.13, we deduce that S is a complete intersection. $\square$

The converse is also true, as we prove next.

Theorem 9.10. *A numerical semigroup other than* $\mathbb{N}$ *is a complete intersection if and only if it is a gluing of two complete intersection numerical semigroups.*

Proof. Sufficiency is Proposition 9.9, so we focus on the necessary condition. Let S be a complete intersection numerical semigroup with minimal system of generators $A = \{n_1, \ldots, n_e\}$. We know that S admits a presentation ρ with $\#\rho = e - 1$. From Lemma 9.5, we deduce that $P_e = \{A\}$, $P_{e-1} = \{A_1, A_2\}$ and that $\rho = \gamma_e = \gamma_{e-1} \cup \{(e_{A_1}(\mu(A_1)), e_{A_2}(\mu(A_2)))\}$, with $\gamma_{e-1} \subseteq \sigma_{A_1} \cup \sigma_{A_2}$. Then A is the gluing of A_1 and A_2. Assume (after rearranging the elements in A if necessary) that $A_1 = \{n_1, \ldots, n_r\}$ and $A_2 = \{n_{r+1}, \ldots, n_e\}$. Let $d_1 = \gcd(A_1)$ and $d_2 = \gcd(A_2)$. Define $S_1 = \langle \frac{n_1}{d_1}, \ldots, \frac{n_r}{d_1} \rangle$ and $S_2 = \langle \frac{n_{r+1}}{d_2}, \ldots, \frac{n_e}{d_2} \rangle$. Clearly $\{\frac{n_1}{d_1}, \ldots, \frac{n_r}{d_1}\}$ and $\{\frac{n_{r+1}}{d_2}, \ldots, \frac{n_e}{d_2}\}$ are minimal systems of generators of S_1 and S_2, respectively. As A is the gluing of A_1 and A_2, by Theorem 9.2, $d_1 d_2 = \mathrm{lcm}\{d_1, d_2\} \in \langle n_1, \ldots, n_r \rangle \cap \langle n_{r+1}, \ldots, n_e \rangle$. Hence $d_2 \in S_1$ and $d_1 \in S_2$. Assume that $d_2 = \frac{n_i}{d_1}$ for some $i \in \{1, \ldots, r\}$. Then $d_1 d_2 = n_i \in \langle n_{r+1}, \ldots, n_e \rangle$, contradicting that A is a minimal system of generators of S. This proves that $d_2 \notin \{n_1, \ldots, n_r\}$. That $d_1 \notin \{n_{r+1}, \ldots, n_e\}$ follows analogously. Hence S is the gluing of S_1 and S_2. Let us prove that S_1 and S_2 are complete intersections. Clearly, $\rho_i = \sigma_{A_i} \cap \rho$ is a presentation of S_i, $i \in \{1, 2\}$. As $\#\rho_1 + \#\rho_2 = e - 2$ and by Theorem 9.6, $\#\rho_1 \geq r - 1$ and $\#\rho_2 \geq e - r - 1$, we conclude that $\#\rho_1 = r - 1$ and that $\#\rho_2 = e - r - 1$. This proves that S_1 and S_2 are complete intersections. $\square$

We end this section proving that every complete intersection numerical semigroup is symmetric. This is a consequence of the following result that shows that the symmetry is preserved under gluing.

Proposition 9.11. *A gluing of symmetric numerical semigroups is symmetric.*

Proof. Let S_1 and S_2 be numerical semigroups with minimal systems of generators $\{n_1, \ldots, n_r\}$ and $\{n_{r+1}, \ldots, n_e\}$, respectively. Let $\lambda \in S_1 \setminus \{n_1, \ldots, n_r\}$ and $\mu \in S_2 \setminus \{n_{r+1}, \ldots, n_e\}$ be such that $\gcd\{\lambda, \mu\} = 1$. Let $S = \langle \mu n_1, \ldots, \mu n_r, \lambda n_{r+1}, \ldots, \lambda n_e \rangle$. By Lemma 9.8, $\{\mu n_1, \ldots, \mu n_r, \lambda n_{r+1}, \ldots, \lambda n_e\}$ is the gluing of $\{\mu n_1, \ldots, \mu n_r\}$ and $\{\lambda n_{r+1}, \ldots, \lambda n_e\}$. By Theorem 9.2 we deduce that

$$\mathrm{Ap}(S, \lambda \mu) = \{s_1 + s_2 \mid s_1 \in \mathrm{Ap}(A_1, \lambda \mu), s_2 \in \mathrm{Ap}(A_2, \lambda \mu)\}.$$

Hence

$$\mathrm{Ap}(S, \lambda \mu) = \{\lambda w_1 + \mu w_2 \mid w_1 \in \mathrm{Ap}(S_1, \lambda), w_2 \in \mathrm{Ap}(S_2, \mu)\}.$$

The result now follows from Proposition 4.10. $\square$

Corollary 9.12. *Every numerical semigroup that is a complete intersection is symmetric.*

4 Free numerical semigroups

In this section we introduce the family of numerical semigroups given by Bertin and Carbonne in [6]. The semigroups in this family are complete intersections. Their definition and characterization rely on the following constants.

Let S be a numerical semigroup minimally generated by $\{n_1, \ldots, n_e\}$. For every $i \in \{2, \ldots, e\}$, define:

1) $d_i = \gcd\{n_1, \ldots, n_{i-1}\}$,
2) $\overline{c_i} = \min\{k \in \mathbb{N} \setminus \{0\} \mid kn_i \text{ is a multiple of } d_i\}$,
3) $c_i^* = \min\{k \in \mathbb{N} \setminus \{0\} \mid kn_i \in \langle n_1, \ldots, n_{i-1}\rangle\}$,
4) $c_i = \min\{k \in \mathbb{N} \setminus \{0\} \mid kn_i \in \langle n_1, \ldots, n_{i-1}, n_{i+1}, \ldots, n_e\rangle\}$.

We study how these constants are related.

Lemma 9.13. *Under the standing hypothesis.*

1) $\overline{c_i} \leq c_i^*$ for all $i \in \{2, \ldots, e\}$,
2) $c_i \leq c_i^*$ for all $i \in \{2, \ldots, e\}$,
3) $\overline{c_i} = \frac{d_i}{d_{i+1}}$ for all $i \in \{2, \ldots, e\}$ (and $d_{e+1} = 1$),
4) $n_1 = \overline{c_2} \cdots \overline{c_e}$.

Proof. 1) and 2) follow directly from the definition. As for 3), note that $\overline{c_i} n_i = \operatorname{lcm}\{n_i, d_i\}$. Since $\operatorname{lcm}\{n_i, d_i\} = \frac{n_i d_i}{\gcd\{n_i, d_i\}} = \frac{n_i d_i}{d_{i+1}}$, we obtain that $\overline{c_i} = \frac{d_i}{d_{i+1}}$. Finally, 4) follows directly applying 3) and the fact that $d_2 = n_1$ and $d_{e+1} = 1$. $\square$

The constants $\overline{c_i}$ can be used to express uniquely any integer as a linear combination of $n_1, \ldots, n_e$.

Lemma 9.14. *Under the standing hypothesis. Every integer z can be written uniquely as $z = \lambda_1 n_1 + \lambda_2 n_2 + \cdots + \lambda_e n_e$, with $\lambda_1 \in \mathbb{Z}$ and $\lambda_i \in \{0, \ldots, \overline{c_i} - 1\}$ for all $i \in \{2, \ldots, p\}$.*

Proof. As $\gcd\{n_1, \ldots, n_e\} = 1$, there exist integers $\mu_1, \ldots, \mu_e$ such that $z = \mu_1 n_1 + \cdots + \mu_e n_e$. By using the division algorithm, there exist integers q and λ_e such that $\mu_e = q\overline{c_e} + \lambda_e$ with $0 \leq \lambda_e < \overline{c_e}$. Hence $z = \mu_1 n_1 + \cdots + \mu_{e-1} n_{e-1} + q\overline{c_e} n_e + \lambda_e n_e$. By substituting $\overline{c_e} n_e$ with its expression in terms of $n_1, \ldots, n_{e-1}$ (via Bézout's identity), we get $z = \gamma_1 n_1 + \cdots + \gamma_{e-1} n_{e-1} + \lambda_e n_e$ with $\gamma_1, \ldots, \gamma_{e-1}$ integers. We repeat the operation with the coefficient of n_{e-1} up to that of n_2, obtaining the desired expression.

Now let us prove that this expression is unique. Assume that $z = \lambda_1 n_1 + \cdots + \lambda_e n_e = \mu_1 n_1 + \cdots + \mu_e n_e$ with $\lambda_1, \ldots, \lambda_e, \mu_1, \ldots, \mu_e$ integers such that $\lambda_i, \mu_i \in \{0, \ldots, \overline{c_i} - 1\}$ for all $i \in \{2, \ldots, p\}$. Let j be the largest integer in $\{1, \ldots, e\}$ such that $\lambda_j \neq \mu_j$. Note that j must be greater than one, since $\lambda_1 n_1 = \mu_1 n_1$ forces λ_1 to be equal to μ_1. Assume without loss of generality that $\lambda_j \geq \mu_j$. Then $(\lambda_j - \mu_j)n_j = (\mu_1 - \lambda_1)n_1 + \cdots + (\mu_{j-1} - \lambda_{j-1})n_{j-1}$, which is a multiple of d_j. As $0 \leq \lambda_j - \mu_j < \overline{c_j}$, from the minimality of $\overline{c_j}$ we deduce that $\lambda_j = \mu_j$, a contradiction. $\square$

The semigroups we are looking for are those for which the relations given in Lemma 9.13 become extreme.

Proposition 9.15. *Under the standing hypothesis, the following are equivalent.*

1) $n_1 = c_2^* \cdots c_e^*$.
2) $\mathrm{Ap}\,(S, n_1) = \{\lambda_2 n_2 + \cdots + \lambda_e n_e \mid \lambda_i \in \{0, \ldots, c_i^* - 1\} \text{ for all } i \in \{2, \ldots, e\}\}$.
3) $\mathrm{F}(S) + n_1 = (c_2^* - 1)n_2 + \cdots + (c_e^* - 1)n_e$.
4) $\overline{c}_i = c_i^*$ for all $i \in \{2, \ldots, e\}$.
5) $c_i = \overline{c}_i$ for all $i \in \{2, \ldots, e\}$.
6) $c_i = \overline{c}_i = c_i^*$ for all $i \in \{2, \ldots, e\}$.

Proof. 1) implies 2). By using the argument of the division algorithm used in Lemma 9.14 but now with c_i^* instead of $\overline{c}_i$, we obtain that every element w of $\mathrm{Ap}(S, n_1)$ admits an expression of the form $w = \lambda_2 n_2 + \cdots + \lambda_e n_e$ with $\lambda_i \in \{0, \ldots, c_i^* - 1\}$. Hence

$$\mathrm{Ap}\,(S, n_1) \subseteq \{\lambda_2 n_2 + \cdots + \lambda_e n_e \mid \lambda_i \in \{0, \ldots, c_i^* - 1\} \text{ for all } i \in \{2, \ldots, e\}\}.$$

The cardinality of this latter set is less than or equal to $c_2^* \cdots c_e^*$. Since by hypothesis $c_2^* \cdots c_e^* = n_1$ and by Lemma 2.4, $\mathrm{Ap}\,(S, n_1) = n_1$, we obtain the desired equality.

2) implies 3). Follows from Proposition 2.12, which asserts that $\mathrm{F}(S) + n_1 = \max \mathrm{Ap}\,(S, n_1)$.

3) implies 4). Arguing as in Lemma 9.14, we can express $\overline{c}_i n_i$ as $\overline{c}_i n_i = \lambda_1 n_1 + \cdots + \lambda_{i-1} n_{i-1}$ with $\lambda_1, \ldots, \lambda_{i-1}$ integers such that $0 \leq \lambda_j < \overline{c}_j$ for all $j \in \{2, \ldots, i - 1\}$. Since $\overline{c}_i \leq c_i^*$, if we prove that $\lambda_1 \in \mathbb{N}$, then $c_i^* = \overline{c}_i$. Assume to the contrary that $\lambda_1 < 0$. Then $\lambda_2 n_2 + \cdots + \lambda_{i-1} n_{i-1} = \overline{c}_i n_i + (-\lambda_1) n_1$. This implies that $\lambda_2 n_2 + \cdots + \lambda_{i-1} n_{i-1} \notin \mathrm{Ap}\,(S, n_1)$ and consequently $(c_2^* - 1)n_2 + \cdots + (c_e^* - 1)n_e \notin \mathrm{Ap}\,(S, n_1)$. This, in view of Proposition 2.12, contradicts 3).

4) implies 5). By definition, $c_e = c_e^*$. Thus $c_e = \overline{c}_e$. Assume that $\overline{c}_{j+1} = c_{j+1}$, $\ldots$, $\overline{c}_e = c_e$ and let us prove that $\overline{c}_j = c_j$. From the definition of c_i^*, by using once more the division algorithm procedure, we deduce that $c_j n_j$ can be expressed as $c_j n_j = a_1 n_1 + \cdots + a_{j-1} n_{j-1} + a_{j+1} n_{j+1} + \cdots + a_e n_e$ with $a_1, \ldots, a_{j-1}, a_{j+1}, \ldots, a_e \in \mathbb{N}$ and $a_{j+k} < c_{j+k}^*$ for all $j + k \in \{j + 1, \ldots, e\}$. Since $\overline{c}_e = c_e^*$, we have that a_e must be zero, since otherwise $a_e n_e = -a_1 n_1 + \cdots + (-a_{j-1})n_{j-1} + c_j n_j + (-a_{j+1})n_{j+1} + \cdots + (-a_{e-1})n_{e-1}$, which is a multiple of d_e and $a_e < \overline{c}_e$. By repeating this argument, we obtain that $a_{j+1} = \cdots = a_e = 0$. Hence $c_j = c_j^*$ and consequently $c_j = \overline{c}_j$.

5) implies 6). Let us see that $c_i = c_i^*$ for all $i \in \{2, \ldots, e\}$. Clearly $c_e = c_e^*$. Now assume that $c_{j+1} = c_{j+1}^*, \ldots, c_e = c_e^*$ and let us prove that $c_j = c_j^*$. In this setting, $\overline{c}_{j+k} = c_{j+k} = c_{j+k}^*$ for all $j + k \in \{j + 1, \ldots, e\}$. Thus we can repeat the steps of the preceding implication.

6) implies 1). This is a direct consequence of Lemma 9.13. $\qquad\square$

A numerical semigroup S is *free* (not to be confused with the notion of free monoid) if there is an arrangement of its set of minimal generators $\{n_1, \ldots, n_e\}$ fulfilling any of the conditions of the last proposition. As mentioned above, free semigroups are complete intersections. This follows easily from the next property.

Theorem 9.16. *Let S be a numerical semigroup other than $\mathbb{N}$. Then S is free if and only if S is a gluing of a free numerical semigroup with embedding dimension $e(S) - 1$ and $\mathbb{N}$.*

Proof. Necessity. Let $\{n_1,\ldots,n_e\}$ be a minimal system of generators arranged so that it fulfills the conditions of Proposition 9.15. We prove that S is a gluing of $T = \langle \frac{n_1}{d_e},\ldots,\frac{n_{e-1}}{d_e}\rangle$ and $\mathbb{N} = \langle 1 \rangle$ (recall that $d_e = \gcd\{n_1,\ldots,n_{e-1}\}$). As $\gcd\{n_e,d_e\} = 1$, $d_e = \overline{c_e}$, which by hypothesis equals $c_e \geq 2$. Thus

$$d_e n_e \in \langle n_1,\ldots,n_{e-1}\rangle \setminus \{n_1,\ldots,n_{e-1}\}.$$

Hence $n_e \in T \setminus \{\frac{n_1}{d_e},\ldots,\frac{n_{e-1}}{d_e}\}$. Since $\{n_1,\ldots,n_e\} = \{d_e\frac{n_1}{d_e},\ldots,d_e\frac{n_{e-1}}{d_e}, n_e \cdot 1\}$, we deduce that S is the gluing of T and $\mathbb{N}$. Now we prove that T is free. Define

$$\hat{c}_i^* = \min\left\{ k \in \mathbb{N}\setminus\{0\} \;\Big|\; k\frac{n_i}{d_e} \in \left\langle \frac{n_1}{d_e},\ldots,\frac{n_{i-1}}{d_e}\right\rangle \right\}.$$

For $i \in \{2,\ldots,e-1\}$, clearly $\hat{c}_i^* = c_i^*$. As $d_e = c_e^*$, $\hat{c}_e^* = d_e$. Since S is free, we know that $n_1 = c_2^* \cdots c_e^*$, whence $\frac{n_1}{d_e} = \hat{c}_2^* \cdots \hat{c}_{e-1}^*$. This proves that T is free in view of Proposition 9.15.

Sufficiency. Assume that S is a gluing of T, minimally generated by $\{m_1,\ldots,m_{e-1}\}$, and $\mathbb{N} = \langle 1 \rangle$. Then there exists a minimal system of generators of S of the form $\{n_1,\ldots,n_e\} = \{dm_1,\ldots,dm_{e-1}, n_e \cdot 1\}$ for some $d \in \mathbb{N}\setminus\{1\}$ and some $n_e \in T \setminus \{m_1,\ldots,m_{e-1}\}$. Moreover, $dn_e \in \langle dm_1,\ldots dm_{e-1}\rangle \cap \langle n_e \rangle$. Define c_i^* and $\hat{c}_i^*$ as above. Observe that any element in $\langle n_1,\ldots,n_{e-1}\rangle$ is a multiple of d and that $\gcd\{d,n_e\} = 1$. This implies that $c_e^* = d$. Moreover, $c_i^* = \hat{c}_i^*$ for all $i \in \{2,\ldots,e-1\}$. As T is free, $m_1 = \hat{c}_2^* \cdots \hat{c}_{e-1}^*$. Hence $dm_1 = n_1 = c_2^* \cdots c_{e-1}^* c_e^*$. This proves that S is free because it fulfills one of the conditions of Proposition 9.15. $\qquad\square$

By using induction and that $\mathbb{N}$ is a complete intersection, together with Theorem 9.10, one easily proves the following consequence of this characterization.

Corollary 9.17. *Every free numerical semigroup is a complete intersection.*

As another consequence we get an explicit description of the shape of a minimal presentation for a free numerical semigroup.

Corollary 9.18. *Let S be a numerical semigroup for the arrangement of its minimal set of generators $\{n_1,\ldots,n_e\}$. Assume that $c_i^* n_i = a_{i_1} n_1 + \cdots + a_{i_{i-1}} n_{i-1}$ for some $a_{i_1},\ldots,a_{i_{i-1}} \in \mathbb{N}$. Then*

$$\left\{ (c_i^* x_i, a_{i_1} x_1 + \cdots + a_{i_{i-1}} x_{i-1}) \mid i \in \{2,\ldots,e\} \right\}$$

is a minimal presentation of S.

Proof. Follows by induction by using Theorem 9.16, Lemma 9.9 and the definition of gluing. $\qquad\square$

Moreover, free numerical semigroups can be characterized as those having a minimal presentation with the stair shape of the presentation given in this last corollary.

Corollary 9.19. *Let S be a numerical semigroup. The following are equivalent.*

1) S is free for the arrangement of its minimal generators $\{n_1,\ldots,n_e\}$.
2) S admits a minimal presentation of the form

$$\left\{ (a_i x_i, a_{i_1} x_1 + \cdots + a_{i_{i-1}} x_{i-1}) \mid i \in \{2,\ldots,e\} \right\}.$$

Proof. 1) implies 2). This is a consequence of Corollary 9.18.

2) implies 1). We use induction on e. For $e = 1$ or $e = 2$ the result is true. Assume that the result holds for $e - 1$. From the hypothesis, we deduce that $\{n_1,\ldots,n_p\}$ is a gluing of $\{n_1,\ldots,n_{e-1}\}$ and $\{n_e\}$. By induction hypothesis, and setting $d = \gcd\{n_1,\ldots,n_{e-1}\}$, we deduce that $T = \langle \frac{n_1}{d},\ldots,\frac{n_{e-1}}{d}\rangle$ is free. Finally, by using Theorem 9.16 we deduce that S is free. $\qquad\qquad\square$

Example 9.20. Let S be a numerical semigroup minimally generated by $\{n_1,n_2,n_3\}$. If the cardinality of $\{c_1 n_1, c_2 n_2, c_3 n_3\}$ is less than three, then by the above corollary and Example 8.23, S is free.

Exercises

Exercise 9.1. Let S be a numerical semigroup with embedding dimension greater than two. Assume that every two minimal generators of S are coprime. Prove that S is not a complete intersection.

Exercise 9.2. Prove that for every integer e greater than one, there exists a complete intersection numerical semigroup with embedding dimension e.

Exercise 9.3. Let S be a symmetric numerical semigroup with $e(S) \in \{m(S) - 1, m(S) - 2, m(S) - 3\}$. Prove that S is a complete intersection if and only if $e(S) = 3$ (see Exercise 8.8).

Exercise 9.4. Find a numerical semigroup with embedding dimension 6 and such that the cardinality of any of its minimal presentations is 7.

Exercise 9.5. Show with an example that the gluing of two free numerical semigroups needs not to be free.

Exercise 9.6. Let S be a numerical semigroup with maximal embedding dimension. Prove that S is a complete intersection if and only if $e(S) \in \{1,2\}$.

Exercise 9.7 ([33]). Let a and b be integers with $1 \le b < a$ and let $S = \langle a, a + 1,\ldots,a+b\rangle$. Prove that S is a complete intersection if and only if $b = 1$, or $b = 2$ and a is even.

Exercise 9.8 ([72]). Let $x_1, \ldots, x_p, y_1, \ldots, y_p$ be integers greater than one with

$$\gcd\{y_1 \cdots y_i, x_i\} = 1 \quad \text{for all } i \in \{1, \ldots, p\}.$$

Prove that

$$S = \langle x_1 \cdots x_p, y_1 x_2 \cdots x_p, \ldots, y_1 \cdots y_{p-1} x_p, y_1 \cdots y_p \rangle$$

is a free semigroup with $e(S) = p + 1$.

Exercise 9.9 ([72]). Let $x_1, \ldots, x_p, y_1, \ldots, y_p$ be integers greater than one with

$$\gcd\{x_i, x_j\} = 1 \text{ for } i \neq j \quad \text{and} \quad \gcd\{x_i, y_i\} = 1 \text{ for all } i \in \{1, \ldots, p\}.$$

Let $M = x_1 \cdots x_p$. Prove that

$$S = \left\langle M, \frac{y_1}{x_1} M, \ldots, \frac{y_p}{x_p} M \right\rangle$$

is a free semigroup with $e(S) = p + 1$.

Exercise 9.10 ([72]). Let a and b be integers greater than one with $\gcd\{a, b\} = 1$. Prove that for every positive integer p,

$$S = \langle a^p, a^p + b, a^p + ab, \ldots, a^p + a^{p-1} b \rangle$$

is a free semigroup with $e(S) = p + 1$.

Exercise 9.11. Let S be a numerical semigroup. Define $S_{(k)}$ as the set of elements in S that can be expressed as sums of k nonzero elements of S. Prove that $S_{(k)} \cup \{0\}$ is a numerical semigroup.

Exercise 9.12 ([86]). A numerical semigroup S minimally generated by $\{n_1 < \cdots < n_e\}$ is *telescopic* if it is free for the arrangement of generators $n_1, \ldots, n_e$. Let

$$\varphi : \text{Free}(x_1, \ldots, x_e) \to S, \quad \varphi(a_1 x_1 + \cdots + a_e x_e) = a_1 n_1 + \cdots + a_e n_e.$$

Prove that if S is telescopic, then the map φ defines a one to one correspondence between the set

$$\{a_1 x_1 + \cdots + a_e x_e \mid a_1 + \cdots + a_e \geq k, 0 \leq a_i < c_i, i \in \{2, \ldots, k\}\}$$

and $S_{(k)}$.

Chapter 9
Numerical semigroups with embedding dimension three

Introduction

Herzog in [41] proves that for embedding dimension three numerical semigroups, the concepts of symmetric and complete intersection coincide. Hence with the help of Proposition 2.17 and what we know about embedding dimension two numerical semigroups, a formula for the Frobenius number and the genus of a symmetric numerical semigroup with embedding dimension three can easily be found.

As for the pseudo-symmetric case, an expression for the Frobenius number of a numerical semigroup of embedding dimension three can be given in terms of the generators (and consequently also a formula for the genus in view of Corollary 4.5). This formula is presented by the authors in [82].

The general case is not that simple. There is no algebraic formula in terms of the minimal generators ([21]). However, an analytic formula is given in [29]. The algebraic formulas known so far for the Frobenius number of a numerical semigroup minimally generated by $\langle n_1, n_2, n_3 \rangle$ depend on the constants $c_i = \min\{x \in \mathbb{N} \setminus \{0\} \mid x n_i \in \langle \{n_1, n_2, n_3\} \setminus \{n_i\}\rangle\}$ and how $c_i n_i$ is expressed in terms of the other two generators (see [13, 42]). In [81] the authors make an extensive study of these semigroups.

1 Numerical semigroups with Apéry sets of unique expression

We start this chapter by reviewing some results appearing in [61] for a special class of numerical semigroups. These properties will be used later for embedding dimension three numerical semigroups.

Let S be a numerical semigroup minimally generated by $\{n_1 < \cdots < n_e\}$, and let $X = \{x_1, \ldots, x_e\}$. As usual let $\varphi : \mathrm{Free}(X) \to S$ be the map defined by $\varphi(\lambda_1 x_1 + \cdots + \lambda_e x_e) = \lambda_1 n_1 + \cdots + \lambda_e n_e$ with kernel congruence σ. Recall that the set of expressions of $n \in \mathbb{N}$ is defined as $\mathrm{Z}(n) = \varphi^{-1}(n)$. We say that $n \in S$ has

J.C. Rosales, P.A. García-Sánchez, *Numerical Semigroups,*
Developments in Mathematics 20, DOI 10.1007/978-1-4419-0160-6_10,
© Springer Science+Business Media, LLC 2009

unique expression if the set $Z(n)$ has only one element. We say that S is a numerical semigroup with *Apéry set of unique expression* if every element in $\mathrm{Ap}(S, n_1)$ has unique expression.

We see that minimal presentations for semigroups with Apéry set of unique expression are relatively easy to control. Define

$$\mathrm{J}(S) = \{\lambda_2 x_2 + \cdots + \lambda_e x_e \mid \lambda_2 n_2 + \cdots + \lambda_e n_e \notin \mathrm{Ap}(S, n_1)\}.$$

Recall that by Dickson's lemma (Lemma 8.6) the set of minimal elements of $\mathrm{J}(S)$ with respect to the ordering $\leq$ is finite. Assume that $\{a_1, \ldots, a_t\} = \mathrm{Minimals}_{\leq}(\mathrm{J}(S))$. For all $i \in \{1, \ldots, t\}$, as $\varphi(a_i) \notin \mathrm{Ap}(S, n_1)$, there exists $b_i \in Z(\varphi(a_i))$ such that $x_1 \leq b_i$. For $i \in \{1, \ldots, t\}$, write

$$a_i = a_{i_2} x_2 + \cdots + a_{i_e} x_e, \ b_i = b_{i_1} x_1 + \cdots + b_{i_e} x_e$$

(note that $b_{i_1} \neq 0$).

These elements can be used to compute a minimal presentation for S, if S has Apéry set of unique expression. First we see what are the R-classes of a_i.

Lemma 10.1. *Under the standing hypothesis, if S is a numerical semigroup with Apéry set of unique expression, then for all $i \in \{1, \ldots, t\}$, the R-class of $Z(\varphi(a_i))$ containing a_i has just one element, that is, $[a_i]_R = \{a_i\}$.*

Proof. Let $n = \varphi(a_i)$. Assume to the contrary that there exists $c \in Z(n)$, $c \neq a_i$, such that $a_i \cdot c \neq 0$. Then there exists $j \in \{1, \ldots, e\}$ such that $x_j \leq a_i$ and $x_j \leq c$. As $x_1 \not\leq a_i$, this implies that $j \geq 2$. By the minimality of a_i, the element $a_i - x_j \notin \mathrm{J}(S)$, or in other words, $\varphi(a_i - x_j) \in \mathrm{Ap}(S, n_1)$. But then $\varphi(a_i - x_j)$ admits at least two different expressions, say $a_i - x_j$ and $c - x_j$, contradicting that S has Apéry set of unique expression. $\qquad\qquad\square$

Theorem 10.2. *Let S be a numerical semigroup with Apéry set of unique expression. Let $\{a_1, \ldots, a_t\} = \mathrm{Minimals}_{\leq}\mathrm{J}(S)$ and let $b_1, \ldots, b_t$ be such that $x_1 \leq b_i$ and $\varphi(a_i) = \varphi(b_i)$ for all $i \in \{1, \ldots, t\}$. Then*

$$\{(a_1, b_1), \ldots, (a_t, b_t)\}$$

is a minimal presentation of S.

Proof. Let $\rho = \{(a_1, b_1), \ldots, (a_t, b_t)\}$. In order to see that ρ is a minimal presentation of S, we see that it can be obtained by following the construction given in Remark 8.12. By definition, $\rho \subset \sigma$, and thus $\rho = \bigcup_{n \in S \setminus \{0\}} \rho \cap (Z(n) \times Z(n))$. From the preceding lemma, we deduce that a_i and b_i are in different R-classes of $Z(\varphi(a_i))$. Hence if $n \in S$ is such that $Z(n)$ has a unique R-class, then $\rho \cap (Z(n) \times Z(n))$ is empty.

Assume that n is an element in S such that the R-classes of $Z(n)$ are $X_1, \ldots, X_r$, with $r \geq 2$. As S has Apéry set with unique expression and $r \geq 2$, we have that $n \notin \mathrm{Ap}(S, n_1)$. Hence there exists at least an element $b \in Z(n)$ such that $x_1 \leq b$.

Assume without loss of generality that X_1 contains all the elements in $Z(n)$ that are greater than x_1 (with respect to $\leq$). In order to conclude the proof, it suffices to prove that if $a \in X_i$, with $i \geq 2$, then $a \in \{a_1, \ldots, a_t\}$. Since $a \notin A_1$, $a = \lambda_2 x_2 + \cdots + \lambda_e x_e$ for some $\lambda_2, \ldots, \lambda_e \in \mathbb{N}$. We already know that $n \notin \mathrm{Ap}(S, n_1)$, and consequently $a \in J(S)$. Moreover, let $j \in \{2, \ldots, e\}$ be such that $\lambda_j \neq 0$. If $n - n_j = \lambda_2 n_2 + \cdots + (\lambda_j - 1)n_j + \cdots + \lambda_e n_e \notin \mathrm{Ap}(S, n_1)$, then $n - (n_j + n_1) \in S$, which contradicts Lemma 8.15. Hence $\lambda_2 n_2 + \cdots + (\lambda_j - 1)n_j + \cdots + \lambda_e n_e \in \mathrm{Ap}(S, n_1)$, which means that $\lambda_2 x_2 + \cdots + (\lambda_j - 1)x_j + \cdots + \lambda_e x_e \notin J(S)$. This proves that $a \in \mathrm{Minimals}_{\leq} J(S)$. $\square$

We now characterize numerical semigroups with Apéry set of unique expression that are symmetric. These numerical semigroups turn out to be free.

Recall that a numerical semigroup S is symmetric if and only if $\mathrm{Maximals}_{\leq_S} \mathrm{Ap}(S, \mathrm{m}(S)) = \{F(S) + \mathrm{m}(S)\}$ (see Corollary 4.12). Different expressions of an element w in $\mathrm{Ap}(S, \mathrm{m}(S))$ translate to different expressions of any $w' \in \mathrm{Ap}(S, \mathrm{m}(S))$ such that $w \leq_S w'$. Hence a numerical semigroup has Apéry set of unique expression if and only if the elements in $\mathrm{Maximals}_{\leq_S}(\mathrm{Ap}(S, \mathrm{m}(S)))$ have unique expressions. By Corollary 4.12, a symmetric numerical semigroup has Apéry set of unique expression if and only if the element $F(S) + \mathrm{m}(S)$ is of unique expression.

Lemma 10.3. *Let S be a symmetric numerical semigroup. Then S has Apéry set of unique expression if and only if $F(S) + \mathrm{m}(S)$ has unique expression.*

We prove now that free and symmetric properties coincide for numerical semigroups with Apéry set of unique expression.

Proposition 10.4. *Let S be a numerical semigroup with Apéry set of unique expression. Then S is symmetric if and only if it is free.*

Proof. We already know that if S is free, then it is a complete intersection (Corollary 9.17), and thus it is symmetric (Corollary 9.12).

Now assume that S is symmetric and that $\{n_1, \ldots, n_e\}$ is a minimal system of generators of S. Suppose that $n_1 = \min\{n_1, \ldots, n_e\}$. For every $j \in \{1, \ldots, e\}$, define $c_i = \min\{k \in \mathbb{N} \setminus \{0\} \mid kn_j \in \langle n_1, \ldots, n_{j-1}, n_{j+1}, \ldots, n_e \rangle\}$. Assume that $c_j n_j = \lambda_{j_1} n_1 + \cdots + \lambda_{j_e} n_e$ for some $\lambda_{j_1}, \ldots, \lambda_{j_e} \in \mathbb{N}$ and $\lambda_{j_j} = 0$. Then, as S has Apéry set of unique expression, $c_j n_j \notin \mathrm{Ap}(S, n_1)$, and thus we can take λ_{j_1} to be nonzero. Moreover, from the minimality of c_j, we deduce that $(c_j - 1)n_j \in \mathrm{Ap}(S, n_1)$. As S is symmetric, we know that $\{F(S) + n_1\} = \mathrm{Maximals}_{\leq_S} \mathrm{Ap}(S, n_1)$ (Corollary 4.12). Hence for all $j \in \{2, \ldots, e\}$, there exists $s_j \in \mathrm{Ap}(S, n_1)$ such that $F(S) + n_1 = (c_j - 1)n_j + s_j$. As $F(S) + n_1$ has unique expression and $c_j n_j \notin \mathrm{Ap}(S, n_1)$, we deduce that $F(S) + n_1 = (c_2 - 1)n_2 + \cdots + (c_e - 1)n_e$. Note that this also proves that

$$\mathrm{Ap}(S, n_1) = \{\lambda_2 n_2 + \cdots + \lambda_e n_e \mid \lambda_i \in \{0, \ldots, c_j - 1\} \text{ for all } j \in \{2, \ldots, e\}\}.$$

From the equality

$$(c_2 - 1)n_2 + \cdots + (c_e - 1)n_e = (\lambda_{2_1} + \cdots + \lambda_{e_1})n_1$$
$$+ (\lambda_{2_2} + \cdots + \lambda_{e_2} - 1)n_2 + \cdots + (\lambda_{2_e} + \cdots + \lambda_{e_e} - 1)n_e,$$

by using that $\lambda_{2_1} + \cdots + \lambda_{e_1} \neq 0$ and that $F(S) + n_1 \in \mathrm{Ap}\,(S, n_1)$, it follows that some of the coefficients in the right-hand side cannot be nonnegative integers. Then there exists $j \in \{2, \ldots, e\}$ such that $\lambda_{2_j} + \cdots + \lambda_{e_j} = 0$. As $\lambda_{2_j}, \ldots, \lambda_{e_j} \in \mathbb{N}$, we have that $\lambda_{2_j} = \cdots = \lambda_{e_j} = 0$. Assume without loss of generality that $j = e$. By iterating the process we obtain that

$$c_2 n_2 = \lambda_{2_1} n_1,$$
$$c_3 n_3 = \lambda_{3_1} n_1 + \lambda_{3_2} n_2,$$
$$\cdots$$
$$c_e n_e = \lambda_{e_1} n_1 + \cdots + \lambda_{e_{e-1}} n_{e-1}.$$

This shows that $c_j = c_j^*$ for all $j \in \{2, \ldots, p\}$, and as $F(S) + n_1 = (c_2 - 1)n_2 + \cdots + (c_e - 1)n_e$, we also have that

$$\mathrm{Ap}\,(S, n_1) = \left\{\, \lambda_2 n_2 + \cdots + \lambda_e n_e \mid \lambda_i \in \{0, \ldots, c_j^* - 1\} \text{ for all } j \in \{2, \ldots, e\} \,\right\}.$$

This proves, by using Proposition 9.15, that S is free.   □

For embedding dimension three numerical semigroups a bit more is achieved, since we now prove that the complete intersection, symmetric and free properties coincide in this scope.

Corollary 10.5. *Let S be a numerical semigroup with embedding dimension three. The following are equivalent.*

1) S is a complete intersection.
2) S is symmetric.
3) S is free.

Proof. We know by Example 9.20 and by Corollary 9.17 that for numerical semigroups of embedding dimension three, it is equivalent to be free and a complete intersection. We also know that every complete intersection is symmetric (Corollary 9.12). So it suffices to prove that if S is symmetric, then it is free. Let S be a symmetric numerical semigroup minimally generated by $\{n_1 < n_2 < n_3\}$. If S has Apéry set of unique expression, then by the preceding proposition, we are done. Hence assume that there exists $w \in \mathrm{Ap}\,(S, n_1)$ and $\lambda_2, \lambda_3, \mu_2, \mu_3 \in \mathbb{N}$ such that $w = \lambda_2 n_2 + \lambda_3 n_3 = \mu_2 n_2 + \mu_3 n_3$, with $(\lambda_2, \lambda_3) \neq (\mu_2, \mu_3)$. Assume without loss of generality that $\lambda_2 > \mu_2$. Then $(\lambda_2 - \mu_2)n_2 = (\mu_3 - \lambda_3)n_3 \in \mathrm{Ap}\,(S, n_1)$. Let $\gamma_1, \gamma_3 \in \mathbb{N}$ be such that $c_2 n_2 = \gamma_1 n_1 + \gamma_3 n_3$, with as usual $c_2 = \min\{k \in \mathbb{N} \setminus \{0\} \mid k n_2 \in \langle n_1, n_3 \rangle\}$. Then by definition, $c_2 \leq (\lambda_2 - \mu_2)$. Let $k_2 = c_2 - (\lambda_2 - \mu_2)$. We have that $c_2 n_2 = (\lambda_2 - \mu_2 + k_2)n_2 = k_2 n_2 + (\mu_3 - \lambda_3)n_3$. From the minimality of c_2 we deduce that $k_2 = 0$. This proves that $c_2 = \lambda_2 - \mu_2$, and in the same way we can show that $c_3 = \mu_3 - \lambda_3$. Thus $c_2 n_2 = c_3 n_3$. By Example 9.20 we have that S is free.   □

2 Irreducible numerical semigroups with embedding dimension three

We already know that symmetric numerical semigroups with embedding dimension three are free, and thus they must be a gluing of a numerical semigroup with embedding dimension two and $\mathbb{N}$. Hence we are going to be able to describe the generators of these semigroups, and also we can give explicit expressions of their Frobenius numbers and genus. Pseudo-symmetric numerical semigroups of embedding dimension three cannot be a gluing as happens with the symmetric case. However, by using some tricks we will be able to parameterize the minimal generators of this family of numerical semigroups and will also find closed expressions for the Frobenius number and genus of these semigroups.

2.1 The symmetric case

We already know that symmetric numerical semigroups with embedding dimension three are free and a gluing of a numerical semigroup of embedding dimension two and $\mathbb{N}$. This can be used to give an explicit description of the minimal generators of a semigroup of this kind.

Theorem 10.6. *Let m_1 and m_2 be two relatively prime integers greater than one. Let a, b and c be nonnegative integers with $a \geq 2$, $b+c \geq 2$ and $\gcd\{a, bm_1 + cm_2\} = 1$. Then $S = \langle am_1, am_2, bm_1 + cm_2 \rangle$ is a symmetric numerical semigroup with embedding dimension three. Moreover, every embedding dimension three symmetric numerical semigroup is of this form.*

Proof. Since $\gcd\{am_1, am_2, bm_1 + cm_2\} = \gcd\{a, bm_1 + cm_2\} = 1$, we have that S is a numerical semigroup.

We now prove that $\{am_1, am_2, bm_1 + cm_2\}$ is a minimal system of generators. The elements of $\langle am_1, am_2 \rangle$ are multiples of a, and consequently $bm_1 + cm_2$ cannot belong to this monoid. If $am_1 \in \langle am_2, bm_1 + cm_2 \rangle$, then $am_1 = \lambda am_2 + \mu(bm_1 + cm_2)$ for some $\lambda, \mu \in \mathbb{N}$. This implies that a divides μ, since $\gcd\{a, bm_1 + cm_2\} = 1$. Hence $m_1 = \lambda m_2 + \frac{\mu}{a}(bm_1 + cm_2)$. Assume that $\mu \neq 0$. If $b \neq 0$, then $b = 1$ and $c = 0$, contradicting $b + c \geq 2$. If $b = 0$, then $m_1 = (\lambda + \frac{\mu c}{a})m_2$, contradicting that m_1 and m_2 are relatively prime. Hence $\mu = 0$, and then $m_1 = \lambda m_2$, which again contradicts that m_1 and m_2 are relatively prime. This proves that $am_1 \notin \langle am_2, bm_1 + cm_2 \rangle$. In the same way it can be shown that $am_2 \notin \langle am_1, bm_1 + cm_2 \rangle$.

Finally, observe that S is a gluing of $\langle m_1, m_2 \rangle$ and $\mathbb{N}$. Thus it is a gluing of two symmetric numerical semigroups ($\langle m_1, m_2 \rangle$ is symmetric by Corollary 4.7, and $\mathbb{N}$ is trivially symmetric). Proposition 9.11 ensures that S is symmetric.

Assume now that S is an embedding dimension three symmetric numerical semigroup. By Corollary 10.5, we know that S is free. Assume that it is free for the arrangement $\{n_1, n_2, n_3\}$ of its minimal generators. By Proposition 9.15, we have that $\overline{c_3}n_3 \in \langle n_1, n_2 \rangle$, and by Lemma 9.13, we have that $\overline{c_3} = \gcd\{n_1, n_2\}$. Let b and c be nonnegative integers such that $\overline{c_3}n_3 = bn_1 + cn_2$. Note that $b + c \geq 2$ and $\overline{c_3} \geq 2$, since otherwise $\{n_1, n_2, n_3\}$ would not be a minimal system of generators. By setting $a = \overline{c_3}$, $m_1 = \frac{n_1}{a}$ and $m_2 = \frac{n_2}{a}$, we obtain the desired result. $\square$

Remark 10.7. Assume that a, b, c, m_1 and m_2 are as in Theorem 10.6. From Proposition 2.17, it follows that $F(\langle am_1, am_2, bm_1 + cm_2 \rangle) = aF(\langle m_1, m_2 \rangle) + (a-1)(bm_1 + cm_2)$. Hence

$$F(\langle am_1, bm_2, bm_1 + cm_2 \rangle) = a(m_1m_2 - m_1 - m_2) + (a-1)(bm_1 + cm_2).$$

2.2 The pseudo-symmetric case

In this section we study pseudo-symmetric numerical semigroups with embedding dimension three. These semigroups have been characterized in [82], where an explicit description of the Apéry sets of their multiplicities and the Frobenius number is also presented. We will recall here some of the results appearing in that manuscript.

In this section, S denotes a numerical semigroup minimally generated by $\{n_1, n_2, n_3\}$. Define as usual for $i \in \{1, 2, 3\}$,

$$c_i = \min\{k \in \mathbb{N} \setminus \{0\} \mid kn_i \in \langle \{n_1, n_2, n_3\} \setminus \{n_i\} \rangle\}.$$

Lemma 10.8. *If* $an_i = bn_j + n_k$ *with* $\{n_i, n_j, n_k\} = \{n_1, n_2, n_3\}$, $a, b \in \mathbb{N}$ *and* $b < c_j$, *then* $a = c_i$.

Proof. Let $q, r \in \mathbb{N}$ be such that $a = qc_i + r$ with $0 \leq r < c_i$. From the definition of c_i, there exists $\lambda, \mu \in \mathbb{N}$ such that $c_in_i = \lambda n_j + \mu n_k$. As $an_i = bn_j + n_k$, we deduce that $q\lambda n_j + q\mu n_k + rn_i = bn_j + n_k$. If $\mu = 0$, then $q\lambda n_j + rn_i = bn_j + n_k$. By assumption $\{n_1, n_2, n_3\}$ is the minimal system of generators of S, which implies that $b > q\lambda$. Hence $rn_i = (b - q\lambda)n_j + n_k$, and as $r < c_i$, this contradicts the definition of c_i. Thus $\mu \neq 0$. If $q = 0$, then we get the same contradiction. Hence $q\mu > 0$ and $bn_j = q\lambda n_j + (q\mu - 1)n_k + rn_i$, which leads to $b \geq q\lambda$ and $(b - q\lambda)n_j = (q\mu - 1)n_k + rn_i$. By hypothesis $b < c_j$, and consequently $b - q\lambda = 0$. This implies that $r = 0$ and $q\mu = 1$. This yields $\mu = q = 1$. We conclude that $a = c_i$. $\square$

Recall (Corollary 4.19) that $S = \langle n_1, n_2, n_3 \rangle$ is pseudo-symmetric if and only if $\text{Maximals}_{\leq_S}(\text{Ap}(S, n_1)) = \{F(S) + n_1, \frac{F(S)}{2} + n_1\}$. We are going to see what these two elements are. First we focus on $\frac{F(S)}{2} + n_1$, discarding a possible value that leads to the symmetric case. The results we obtain are enounced for n_1 but can analogously be obtained for n_2 and n_3.

Lemma 10.9. *If* $F(S) + n_1 = (c_2 - 1)n_2 + (c_3 - 1)n_3$, *then* S *is symmetric.*

Proof. Assume that $c_2n_2 \in \mathrm{Ap}(S, n_1)$. Then $c_2n_2 = an_3$ for some $a \in \mathbb{N}$. This forces a to be equal to c_3 (see the argument in the proof of Corollary 10.5). In view of Example 8.23, this implies that S is a complete intersection and thus it is symmetric (Corollary 9.12 or 10.5). Therefore we next assume that neither c_2n_2 nor c_3n_3 belong to $\mathrm{Ap}(S, n_1)$. In this setting, if $w \in \mathrm{Ap}(S, n_1)$, then $w = an_2 + bn_3$ with $a, b \in \mathbb{N}$, $a < c_2$ and $b < c_3$. By hypothesis $\mathrm{F}(S) + n_1 = (c_2 - 1)n_2 + (c_3 - 1)n_3$. Hence $w \leq_S \mathrm{F}(S) + n_1$. This proves that S is symmetric by Corollary 4.12. $\qquad\square$

There are two possible values for $\frac{\mathrm{F}(S)}{2} + n_1$ as we see next.

Lemma 10.10. *If S is pseudo-symmetric, then $\frac{\mathrm{F}(S)}{2} + n_1 \in \{(c_2 - 1)n_2, (c_3 - 1)n_3\}$.*

Proof. Since S is pseudo-symmetric, arguing as in the proof of Lemma 10.9, we have that neither c_2n_2 nor c_3n_3 belong to $\mathrm{Ap}(S, n_1)$. Assume that $\frac{\mathrm{F}(S)}{2} + n_1 \notin \{(c_2 - 1)n_2, (c_3 - 1)n_3\}$. From the definition of c_2, it follows that $(c_2 - 1)n_2 - n_1 \notin S$. Then by Proposition 4.4, $\mathrm{F}(S) - ((c_2 - 1)n_2 - n_1) = \mathrm{F}(S) + n_1 - (c_2 - 1)n_2 \in S$. As $c_2n_2, c_3n_3 \notin \mathrm{Ap}(S, n_1)$ and $\mathrm{F}(S) + n_1 \in \mathrm{Ap}(S, n_1)$, we have that $\mathrm{F}(S) + n_1 = (c_2 - 1)n_2 + bn_3$ with $b < c_3$. By using now the definition of c_3, we have that $(c_3 - 1)n_3 \in \mathrm{Ap}(S, n_1)$, and since $(c_3 - 1)n_3 \neq \frac{\mathrm{F}(S)}{2} + n_1$, Proposition 4.4 asserts that $\mathrm{F}(S) + n_1 - (c_3 - 1)n_3 \in S$. Arguing as above, there exists $a \in \mathbb{N}$ such that $\mathrm{F}(S) + n_1 = (c_3 - 1)n_3 + an_2$ with $a < c_2$. Hence $an_2 + (c_3 - 1)n_3 = (c_2 - 1)n_2 + bn_3$, $a \leq c_2 - 1$ and $b \leq c_3 - 1$. By using again the definitions of c_2 and c_3, we deduce that $a = c_2 - 1$ and $b = c_3 - 1$. Therefore $\mathrm{F}(S) + n_1 = (c_2 - 1)n_2 + (c_3 - 1)n_3$, which in view of Lemma 10.9, contradicts the fact that S is pseudo-symmetric. $\qquad\square$

This, as we see next, provides us with an explicit description of the Apéry set of any of the minimal generators in a pseudo-symmetric numerical semigroup with embedding dimension three. In particular, in view of Proposition 2.12, this description is telling us what $\mathrm{F}(S) + n_1$ is.

Lemma 10.11. *If S is pseudo-symmetric and $\frac{\mathrm{F}(S)}{2} + n_1 = (c_2 - 1)n_2$, then*

$$\mathrm{Ap}(S, n_1) = \{an_2 + bn_3 \mid 0 \leq a \leq c_2 - 2, \ 0 \leq b \leq c_3 - 1\} \cup \{(c_2 - 1)n_2\}.$$

Proof. We already know that $c_3n_3 \notin \mathrm{Ap}(S, n_1)$ (see the proof of Lemma 10.9). From Corollary 4.19, we deduce that if $a, b \in \mathbb{N}$ and $an_2 + bn_3 \in \mathrm{Ap}(S, n_1) \setminus \{\frac{\mathrm{F}(S)}{2} + n_1\}$, then $a \leq c_2 - 2$ and $b \leq c_3 - 1$. In order to conclude the proof, it suffices to show that $\mathrm{F}(S) + n_1 = (c_2 - 2)n_2 + (c_3 - 1)n_3$. As $(c_2 - 2)n_2, (c_3 - 1)n_3 \in \mathrm{Ap}(S, n_1) \setminus \{\frac{\mathrm{F}(S)}{2} + n_1\}$, by Corollary 4.19, we obtain that $\mathrm{F}(S) + n_1 = (c_3 - 1)n_3 + an_2 = (c_2 - 2)n_2 + bn_3$ for some $a, b \in \mathbb{N}$ with $a \leq c_2 - 2$ and $b \leq c_3 - 1$. From the equality $(c_3 - 1)n_3 + an_2 = (c_2 - 2)n_2 + bn_3$, we deduce that $a = c_2 - 2$ and $b = c_3 - 1$. $\qquad\square$

The parameters in Example 8.23 can be explicitly computed in the pseudo-symmetric case.

Lemma 10.12. *If S is pseudo-symmetric and $\frac{\mathrm{F}(S)}{2} + n_1 = (c_2 - 1)n_2$, then*

1) $c_1 n_1 = (c_2 - 1)n_2 + n_3$,
2) $c_2 n_2 = (c_3 - 1)n_3 + n_1$,
3) $c_3 n_3 = (c_1 - 1)n_1 + n_2$.

Proof. By Corollary 4.19, we have that $(c_2 - 1)n_2 \in \text{Maximals}_{\leq_S}(\text{Ap}(S, n_1))$. Hence $(c_2 - 1)n_2 + n_3 \notin \text{Ap}(S, n_1)$. Thus there exist $a, b, c \in \mathbb{N}$ with $a \neq 0$ such that $(c_2 - 1)n_2 + n_3 = an_1 + bn_2 + cn_3$. Since by Lemma 10.11, $(c_2 - 2)n_2 + n_3 \in \text{Ap}(S, n_1)$, we obtain that $c = 0$ and $b = 0$. Therefore $an_1 = (c_2 - 1)n_2 + n_3$. By using Lemma 10.8, we deduce that $a = c_1$. This proves 1).

As $\frac{F(S)}{2} + n_3 \neq \frac{F(S)}{2} + n_1 = (c_2 - 1)n_2$, in view of Lemma 10.10, we obtain that $\frac{F(S)}{2} + n_3 = (c_1 - 1)n_1$. Arguing as above, we can deduce that $c_3 n_3 = (c_1 - 1)n_1 + n_2$. Finally, $\frac{F(S)}{2} + n_2 = (c_3 - 1)n_3$ and $c_2 n_2 = (c_3 - 1)n_3 + n_1$.   □

These values of the r_{ij}'s characterize in fact the pseudo-symmetric property as we see in the following result.

Proposition 10.13. *Let S be a numerical semigroup with embedding dimension three. Then S is pseudo-symmetric if and only if for some rearrangement of its generators $\{n_1, n_2, n_3\}$ we have that $c_1 n_1 = (c_2 - 1)n_2 + n_3$, $c_2 n_2 = (c_3 - 1)n_3 + n_1$ and $c_3 n_3 = (c_1 - 1)n_1 + n_2$.*

Proof. Necessity. This is a consequence of Lemmas 10.10 and 10.12.
   Sufficiency. In view of Example 8.23, case 3),

$$\rho = \{(c_1 x_1, (c_2 - 1)x_2 + x_3), (c_2 x_2, x_1 + (c_3 - 1)x_3), (c_3 x_3, (c_1 - 1)x_1 + x_2)\}$$

is a minimal presentation for S. By using Proposition 8.4, we deduce that the element $(c_3 - 1)n_3 + (c_2 - 2)n_2$ is of unique expression. This in particular implies that $(c_3 - 1)n_3 + (c_2 - 2)n_2 \in \text{Ap}(S, n_1)$. By using the same argument, $(c_2 - 1)n_2$ also belongs to $\text{Ap}(S, n_1)$. Observe also that $(c_2 - 1)n_2 + n_2 = (c_3 - 1)n_3 + n_1 \notin \text{Ap}(S, n_1)$ and that $(c_2 - 1)n_2 + n_3 = c_1 n_1 \notin \text{Ap}(S, n_1)$ (and clearly $(c_2 - 1)n_2 + n_1 \notin \text{Ap}(S, n_1)$). Hence $(c_2 - 1)n_2 \in \text{Maximals}_{\leq_S}(\text{Ap}(S, n_1))$. The same stands for $(c_3 - 1)n_3 + (c_2 - 2)n_2$. Now let $an_2 + bn_3$, with $a, b \in \mathbb{N}$, be an element of $\text{Ap}(S, n_1)$. Since $c_3 n_3 = (c_1 - 1)n_1 + n_2$, we have that $b < c_3$; and $c_2 n_2 = (c_3 - 1)n_3 + n_1$ forces $a < c_2$. Finally, the equality $c_1 n_1 = (c_2 - 1)n_2 + n_3$ implies that if $a = c_2 - 1$, then b must be zero. This proves that $\text{Maximals}_{\leq_S}(\text{Ap}(S, n_1)) = \{(c_2 - 1)n_2, (c_2 - 2)n_2 + (c_3 - 1)n_3\}$. In view of Corollary 4.19, in order to show that S is pseudo-symmetric, it suffices to show that $2((c_2 - 1)n_2 - n_1) = (c_3 - 1)n_3 + (c_2 - 2)n_2 - n_1$. This equality holds if and only if $(2c_2 - 2)n_2 = (c_3 - 1)n_3 + (c_2 - 2)n_2 + n_1$. As $(2c_2 - 2)n_2 = (c_2 - 2)n_2 + c_2 n_2 = (c_2 - 2)n_2 + (c_3 - 1)n_3 + n_1$, we get the desired equality.   □

As a consequence of this, we obtain that the minimal generators of an embedding dimension three pseudo-symmetric numerical semigroup are pairwise coprime, in contrast with what happens in the symmetric case (Theorem 10.6).

Corollary 10.14. *If S is a pseudo-symmetric numerical semigroup with embedding dimension three, then its minimal generators are pairwise relatively prime.*

Proof. As S is a numerical semigroup $\gcd(\{n_1,n_2,n_3\}) = 1$. By Proposition 10.13, $n_3 = c_1 n_1 - (c_2 - 1)n_2$, and thus $\gcd(\{n_1,n_2\}) = \gcd(\{n_1,n_2,n_3\}) = 1$. The same stands for $\gcd(\{n_1,n_3\})$ and $\gcd(\{n_2,n_3\})$ in view of the rest of the equalities appearing in Proposition 10.13.   □

By counting the elements in the Apéry sets of the minimal generators in an embedding dimension three pseudo-symmetric numerical semigroup, we obtain an interesting consequence.

Corollary 10.15. *Let S be a pseudo-symmetric numerical semigroup with embedding dimension three. Then there exist $a,b,c \in \mathbb{N} \setminus \{0,1\}$ such that*

$$\{c(b-1)+1, (c-1)a+1, b(a-1)+1\}$$

is the minimal system of generators of S.

Proof. Let $\{n_1,n_2,n_3\}$ be the minimal system of generators of S. Since $\#\mathrm{Ap}\,(S,n_1) = n_1$, in view of Lemma 10.11, we deduce that $n_1 = c_3(c_2 - 1) + 1$. In the same way we can obtain that $n_2 = c_1(c_3 - 1) + 1$ and that $n_3 = c_2(c_1 - 1) + 1$.   □

This property characterizes pseudo-symmetric numerical semigroups with embedding dimension three. It also provides us with a parameterization of the minimal generators of this kind of numerical semigroups.

Proposition 10.16. *Let $a,b,c \in \mathbb{N} \setminus \{0,1\}$ be such that $\gcd(c(b-1)+1, (c-1)a + 1) = 1$. Then*
$$S = \langle c(b-1)+1, (c-1)a+1, b(a-1)+1 \rangle$$

is a pseudo-symmetric numerical semigroup with embedding dimension three and such that
$$\mathrm{F}(S) = 2(a-1)(b-1)(c-1) - 2.$$

Proof. Let $n_1 = c(b-1) + 1$, $n_2 = (c-1)a + 1$ and $n_3 = b(a-1) + 1$. Since $\gcd(n_1,n_2) = 1$, the monoid S is a numerical semigroup. The reader can check that $an_1 = (b-1)n_2 + n_3$, $bn_2 = (c-1)n_3 + n_1$ and $cn_3 = (a-1)n_1 + n_2$. Observe that $\gcd\{n_1,n_2\} = \gcd\{n_2,n_3\} = \gcd\{n_1,n_3\} = 1$. If we prove now that $c_1 = a$, $c_2 = b$ and $c_3 = c$, then $\{n_1,n_2,n_3\}$ is a minimal system of generators of S and by Proposition 10.13 we are done. Assume that there exists $x \in \mathbb{N}$ with $0 < x < a$ and $xn_1 = yn_2 + zn_3$ for some $y,z \in \mathbb{N}$. As $n_1 = bn_2 - (c-1)n_3$, we have that $xn_1 = xbn_2 - x(c-1)n_3$. Hence $yn_2 + zn_3 = xbn_2 - x(c-1)n_3$ and thus $(xb-y)n_2 = (z+x(c-1))n_3$. Since $\gcd\{n_2,n_3\} = 1$, this last equality implies that $xb - y = kn_3$ for some positive integer k ($xb - y \neq 0$, because otherwise $z = -x(c-1)$, which is impossible). Hence $xb \geq n_3$ and consequently $(a-1)b \geq n_3$, in contradiction with $n_3 = (a-1)b + 1$. This proves that $a = c_1$. Analogously one obtains that $c_2 = b$ and $c_3 = c$.

From Lemma 10.11 and Proposition 2.12, $\mathrm{F}(S) + n_1 = (c_3 - 1)n_3 + (c_2 - 2)n_2$, and so $\mathrm{F}(S)$ is equal to $(c-1)(b(a-1)+1) + (b-2)(a(c-1)+1) - (c(b-1)+1) = 2(a-1)(b-1)(c-1) - 2$.   □

By gathering all this information together we obtain the following closed formulas for the c_i's and the Frobenius number.

Theorem 10.17. *Let S be a numerical semigroup with embedding dimension three minimally generated by* $\{n_1, n_2, n_3\}$. *Then S is pseudo-symmetric if and only if for some rearrangement of* $\{n_1, n_2, n_3\}$,

$$\left\{ \frac{(n_1 - n_2 + n_3) + \sqrt{(n_1 + n_2 + n_3)^2 - 4(n_1 n_2 + n_1 n_3 + n_2 n_3 - n_1 n_2 n_3)}}{2n_1}, \right.$$
$$\frac{(n_1 + n_2 - n_3) + \sqrt{(n_1 + n_2 + n_3)^2 - 4(n_1 n_2 + n_1 n_3 + n_2 n_3 - n_1 n_2 n_3)}}{2n_2},$$
$$\left. \frac{(-n_1 + n_2 + n_3) + \sqrt{(n_1 + n_2 + n_3)^2 - 4(n_1 n_2 + n_1 n_3 + n_2 n_3 - n_1 n_2 n_3)}}{2n_3} \right\} \subset \mathbb{N}.$$

If this is the case, then

$$F(S) = -(n_1 + n_2 + n_3) + \sqrt{(n_1 + n_2 + n_3)^2 - 4(n_1 n_2 + n_1 n_3 + n_2 n_3 - n_1 n_2 n_3)},$$

and

$$c_1 = \frac{(n_1 - n_2 + n_3) + \sqrt{(n_1 + n_2 + n_3)^2 - 4(n_1 n_2 + n_1 n_3 + n_2 n_3 - n_1 n_2 n_3)}}{2n_1},$$
$$c_2 = \frac{(n_1 + n_2 - n_3) + \sqrt{(n_1 + n_2 + n_3)^2 - 4(n_1 n_2 + n_1 n_3 + n_2 n_3 - n_1 n_2 n_3)}}{2n_2},$$
$$c_3 = \frac{(-n_1 + n_2 + n_3) + \sqrt{(n_1 + n_2 + n_3)^2 - 4(n_1 n_2 + n_1 n_3 + n_2 n_3 - n_1 n_2 n_3)}}{2n_3}.$$

Proof. Consider the system of equations

$$\begin{cases} n_1 = c(b-1) + 1, \\ n_2 = (c-1)a + 1, \\ n_3 = b(a-1) + 1, \end{cases}$$

with unknowns a, b, c. Set

$$\Delta = (n_1 + n_2 + n_3)^2 - 4(n_1 n_2 + n_1 n_3 + n_2 n_3 - n_1 n_2 n_3).$$

The solutions of the above system are

$$(a, b, c) = \left(\frac{(n_1 - n_2 + n_3) - \sqrt{\Delta}}{2n_1}, \frac{(n_1 + n_2 - n_3) - \sqrt{\Delta}}{2n_2}, \frac{(-n_1 + n_2 + n_3) - \sqrt{\Delta}}{2n_3} \right)$$

and

$$(a, b, c) = \left(\frac{(n_1 - n_2 + n_3) + \sqrt{\Delta}}{2n_1}, \frac{(n_1 + n_2 - n_3) + \sqrt{\Delta}}{2n_2}, \frac{(-n_1 + n_2 + n_3) + \sqrt{\Delta}}{2n_3} \right).$$

Observe that

$$\sqrt{\Delta} = \sqrt{(-n_1 + n_2 + n_3)^2 + 4(n_1 - 1)n_2 n_3} > -n_1 + n_2 + n_3,$$
$$\sqrt{\Delta} = \sqrt{(n_1 - n_2 + n_3)^2 + 4n_1(n_2 - 1)n_3} > n_1 - n_2 + n_3,$$
$$\sqrt{\Delta} = \sqrt{(n_1 + n_2 - n_3)^2 + 4n_1 n_2(n_3 - 1)} > n_1 + n_2 - n_3.$$

Hence all these solutions are real numbers and the only positive solution is the second choice of a, b, c.

Sufficiency. If $\{a, b, c\}$ are integers, then by Theorem 10.16 we have that S is pseudo-symmetric (observe that $a, b, c \geq 2$, since $n_1, n_2, n_3 \neq 1$). The reader can check that the formula given by $F(S)$ is obtained by applying that $F(S) = 2(a - 1)(b - 1)(c - 1) - 2$.

Necessity. If S is pseudo-symmetric, then by Corollary 10.15, the above system of equations has a nonnegative integer solution. $\qquad\square$

3 Pseudo-Frobenius numbers and genus of an embedding dimension three numerical semigroup

In this section, we assume that S is a numerical semigroup minimally generated by $\{n_1, n_2, n_3\}$, and that every two minimal generators are relatively prime. This assumption is not restrictive for the computation of the Frobenius number, pseudo-Frobenius numbers and genus of S. This is due to Proposition 2.17 (for the Frobenius number and genus), and to Lemma 2.16 and Proposition 2.20 (for the computation of the pseudo-Frobenius numbers).

The results of this section are extracted from [81]. As usual define for $i \in \{1, 2, 3\}$,

$$c_i = \min\{k \in \mathbb{N} \setminus \{0\} \mid k n_i \in \langle \{n_1, n_2, n_3\} \setminus \{n_i\} \rangle \}.$$

Hence by using the same notation as in Example 8.23, there exist nonnegative integers $r_{12}, r_{13}, r_{21}, r_{23}, r_{31}, r_{32} \in \mathbb{N}$ such that

$$c_1 n_1 = r_{12} n_2 + r_{13} n_3,$$
$$c_2 n_2 = r_{21} n_1 + r_{23} n_3,$$
$$c_3 n_3 = r_{31} n_1 + r_{32} n_2.$$

We now give some of the properties of these integers and show their relevance in the computations we want to perform in this section.

First we show that under the standing hypothesis, no r_{ij} can be zero.

Lemma 10.18. $r_{12}, r_{13}, r_{21}, r_{23}, r_{31}$ *and* r_{32} *are positive integers.*

Proof. Assume that $r_{13} = 0$. Then $c_1 n_1 = r_{12} n_2$. Since $\gcd\{n_1, n_2\} = 1$ and by definition $c_1 \leq n_2$, we have that $c_1 = n_2$. Besides, as $\gcd\{n_1, n_2\} = 1$, we know that there exists $x \in \{1, \ldots, n_2 - 1\}$ such that $x n_1 \equiv n_3 \bmod n_2$. Hence $x n_1 = n_3 + z n_2$ for some integer z. As $\{n_1, n_2, n_3\}$ is a minimal system of generators of S, $z \in \mathbb{N}$ and consequently $c_1 \leq x < n_2$, in contradiction with $c_1 = n_2$. Analogously, one proves that the rest of the r_{ij} are all positive. $\square$

Next, we prove that the c_i's are determined by the r_{ij}'s.

Lemma 10.19. *For every* $\{i, j, k\} = \{1, 2, 3\}$,

$$c_i = r_{ji} + r_{ki}.$$

Proof. Note that $(r_{13} + r_{23})n_3 = (c_1 - r_{21})n_1 + (c_2 - r_{12})n_2$, whence $r_{13} + r_{23} \geq c_3$. In a similar way we obtain that $r_{21} + r_{31} \geq c_1$ and $r_{12} + r_{32} \geq c_2$. Besides, $c_1 n_1 + c_2 n_2 + c_3 n_3 = (r_{21} + r_{31})n_1 + (r_{12} + r_{32})n_2 + (r_{13} + r_{23})n_3$ and consequently $c_1 = r_{21} + r_{31}$, $c_2 = r_{12} + r_{32}$ and $c_3 = r_{13} + r_{23}$. $\square$

With this we can describe the maximal elements (with respect to $\leq_S$) of the Apéry set of any of the minimal generators. This is of interest to us because by Proposition 2.20 it gives a formula for the pseudo-Frobenius numbers and type of S.

Lemma 10.20. *For every* $\{i, j, k\} = \{1, 2, 3\}$,

$$\mathrm{Maximals}_{\leq_S}(\mathrm{Ap}(S, n_i)) = \{(c_j - 1)n_j + (r_{ik} - 1)n_k, (c_k - 1)n_k + (r_{ij} - 1)n_j\}.$$

Proof. We prove the statement for $i = 1$. The rest of the cases follow analogously.

We first show that $(c_3 - 1)n_3 + (r_{12} - 1)n_2 \in \mathrm{Ap}(S, n_1)$. Assume to the contrary that $(c_3 - 1)n_3 + (r_{12} - 1)n_2 = a_1 n_1 + a_2 n_2 + a_3 n_3$ for some $a_1, a_2, a_3 \in \mathbb{N}$ with $a_1 \neq 0$. From the minimality of c_3 and c_2 (which is greater than r_{12} by Lemma 10.19), we have that $a_3 < c_3 - 1$ and $a_2 < r_{12} - 1$. We deduce that $a_1 n_1 = (c_3 - 1 - a_3)n_3 + (r_{12} - 1 - a_2)n_2$ with $c_3 - 1 - a_3, r_{12} - 1 - a_2 \in \mathbb{N}$ and consequently $a_1 \geq c_1$. Let $q \in \mathbb{N} \setminus \{0\}$ and $0 \leq r < c_1$ be such that $a_1 = qc_1 + r$. Then $(c_3 - 1 - a_3)n_3 + (r_{12} - 1 - a_2)n_2 = rn_1 + qr_{12}n_2 + qr_{13}n_3$. Thus $(c_3 - 1 - a_3 - qr_{13})n_3 = rn_1 + (qr_{12} - r_{12} + 1 + a_2)n_2$ with $r \in \mathbb{N}$ and $qr_{12} - r_{12} + 1 + a_2 \in \mathbb{N} \setminus \{0\}$, in contradiction with the definition of c_3.

In a similar way it follows that $(c_2 - 1)n_2 + (r_{13} - 1)n_3 \in \mathrm{Ap}(S, n_1)$.

Take now $an_2 + bn_3 \in \mathrm{Ap}(S, n_1)$. We show that either $(a, b) \leq (r_{12} - 1, c_3 - 1)$ or $(a, b) \leq (c_2 - 1, r_{13} - 1)$. In view of Lemma 10.18, we deduce that $a < c_2$ and $b < c_3$. If $(a, b) \not\leq (r_{12} - 1, c_3 - 1)$, then $a \geq r_{12}$. Let us prove that in this setting $b < r_{13}$. Assume to the contrary that $b \geq r_{13}$. Then $an_2 + bn_3 = r_{12}n_2 + r_{13}n_3 + (a - r_{12})n_2 + (b - r_{13})n_3 = c_1 n_1 + (a - r_{12})n_2 + (b - r_{13})n_3$, contradicting that $an_2 + bn_3 \in \mathrm{Ap}(S, n_1)$. Hence $(a, b) \leq (c_2 - 1, r_{13} - 1)$. $\square$

From Proposition 2.20 we obtain the following.

Proposition 10.21. *S has type two.*

In particular this implies [32, Theorem 14]. Moreover, with this we also recover [32, Theorem 11].

Corollary 10.22. *A numerical semigroup with embedding dimension three has at most type two.*

Proof. Let S be a numerical semigroup minimally generated by $\{n_1, n_2, n_3\}$. If the three minimal generators are pairwise relatively prime, then the above proposition tells us that the semigroup has type two. Otherwise, two of the minimal generators have greatest common divisor $d \neq 1$ (not the three of them because S is a numerical semigroup). From the reduction given in Lemma 2.16, we have that either the semigroup T in that lemma has embedding dimension two or embedding dimension three. The type of T equals the type of S. In the first case T has type one (Corollaries 4.7 and 4.11), and in the second S has type two. $\qquad\square$

Recall that the c_i's are determined by the r_{ij}'s: By counting the elements in $\mathrm{Ap}(S, n_i)$ as we did in Corollary 10.15, we can express the minimal generators in terms of the r_{ij}'s.

Lemma 10.23.

$$n_1 = r_{12}r_{13} + r_{12}r_{23} + r_{13}r_{32},$$
$$n_2 = r_{13}r_{21} + r_{21}r_{23} + r_{23}r_{31},$$
$$n_3 = r_{12}r_{31} + r_{21}r_{32} + r_{31}r_{32}.$$

Proof. We know by Lemma 2.4 that $\#\mathrm{Ap}(S, n) = n$ for all $n \in S \setminus \{0\}$. Note that from the minimality of c_2 and c_3, if $a_2 n_2 + a_3 n_3 = b_2 n_2 + b_3 n_3$ with $a_i, b_i \in \{0, \ldots, c_i - 1\}$, then $(a_2, a_3) = (b_2, b_3)$. Hence by Lemma 10.20, we deduce that

$$\#\mathrm{Ap}(S, n_1) = \#\left\{(a, b) \in \mathbb{N}^2 \mid (a, b) \leq (r_{12} - 1, c_3 - 1) \text{ or } (a, b) \leq (c_2 - 1, r_{13} - 1)\right\}$$
$$= r_{12}c_3 + c_2 r_{13} - r_{12}r_{13}.$$

Hence $n_1 = r_{12}c_3 + c_2 r_{13} - r_{12}r_{13}$, and by Lemma 10.19, $n_1 = r_{12}r_{13} + r_{12}r_{23} + r_{32}r_{13}$.

The corresponding equalities for n_2 and n_3 follow analogously. $\qquad\square$

We go for the converse. By choosing r_{ij} random and positive, we obtain an embedding dimension three numerical semigroup.

Lemma 10.24. *Let $a_{12}, a_{13}, a_{21}, a_{23}, a_{31}, a_{32}$ be positive integers and let*

$$m_1 = a_{12}a_{13} + a_{12}a_{23} + a_{13}a_{32},$$
$$m_2 = a_{13}a_{21} + a_{21}a_{23} + a_{23}a_{31},$$
$$m_3 = a_{12}a_{31} + a_{21}a_{32} + a_{31}a_{32}.$$

For every $\{i, j, k\} = \{1, 2, 3\}$,

1) $(a_{ji} + a_{ki})m_i = a_{ij}m_j + a_{ik}m_k$,
2) if $\gcd\{m_i, m_j\} = 1$, then $m_k \notin \langle m_i, m_j \rangle$.

Proof. 1) It is easy to check.
2) Let us prove that if $\gcd\{m_1, m_2\} = 1$, then $m_3 \notin \langle m_1, m_2 \rangle$. The other cases follow by symmetry. Assume that $m_3 \in \langle m_1, m_2 \rangle$, then $m_3 = \lambda m_1 + \mu m_2$ for some $\lambda, \mu \in \mathbb{N}$. By 1), we know that $(a_{21} + a_{31})m_1 = a_{12}m_2 + a_{13}m_3$. Hence $a_{13}m_3 = (a_{21} + a_{31})m_1 - a_{12}m_2$. Thus, $a_{13}(\lambda m_1 + \mu m_2) = (a_{21} + a_{31})m_1 - a_{12}m_2$ and $(a_{13}\mu +$

$a_{12})m_2 = (a_{21} + a_{31} - a_{13}\lambda)m_1$. By using that $\gcd\{m_1, m_2\} = 1$, we deduce that $a_{21} + a_{31} - a_{13}\lambda = km_2$ for some $k \in \mathbb{N} \setminus \{0\}$. In particular $a_{21} + a_{31} \geq km_2 \geq m_2$, in contradiction with $m_2 = (a_{21} + a_{31})a_{23} + a_{13}a_{21}$, because $a_{ij} > 0$ for all i, j and consequently $a_{21} + a_{31} < m_2$.   □

Given a sequence of integers $x_1, \ldots, x_p$ we say that it is *strongly positive* if $x_i > 0$ for all $i \in \{1, \ldots, p\}$.

Theorem 10.25. *Let m_1, m_2 and m_3 be positive integers such that $\gcd\{m_i, m_j\} = 1$ for $i \neq j$. Then the system of equations*

$$m_1 = x_{12}x_{13} + x_{12}x_{23} + x_{13}x_{32},$$
$$m_2 = x_{13}x_{21} + x_{21}x_{23} + x_{23}x_{31},$$
$$m_3 = x_{12}x_{31} + x_{21}x_{32} + x_{31}x_{32},$$

has a strongly positive integer solution if and only if $\{m_1, m_2, m_3\}$ is a minimal system of generators of $\langle m_1, m_2, m_3 \rangle$.

Moreover, if such a solution exists, then it is unique.

Proof. Necessity. Follows from Lemma 10.24.

Sufficiency. This is a consequence of Lemmas 10.18 and 10.23.

Now let us prove the uniqueness of the solution. For $n_i = m_i$, $i \in \{1, 2, 3\}$, we know in view of Lemmas 10.18 and 10.23 that $(x_{12}, x_{13}, x_{21}, x_{23}, x_{31}, x_{32}) = (r_{12}, r_{13}, r_{21}, r_{23}, r_{31}, r_{32})$ is a strongly positive integer solution. Assume that $(a_{12}, a_{13}, a_{21}, a_{23}, a_{31}, a_{32})$ is another strongly positive integer solution of the above system of equations. Then we must have that $a_{ij} < r_{ij}$ for some i, j. Without loss of generality we can assume that $a_{12} < r_{12}$. Then $c_1 n_1 = r_{12}n_2 + r_{13}n_3 = (a_{12} + \lambda)n_2 + r_{13}n_3$ for some $\lambda \in \mathbb{N} \setminus \{0\}$. By 1) in Lemma 10.24 we deduce that $a_{12}n_2 = (a_{21} + a_{31})n_1 - a_{13}n_3$ and that $a_{21} + a_{31} \geq c_1$. Hence $c_1 n_1 = (a_{21} + a_{31})n_1 - a_{13}n_3 + \lambda n_2 + r_{13}n_3$. Thus, $(a_{13} - r_{13})n_3 = (a_{21} + a_{31} - c_1)n_1 + \lambda n_2$ and consequently $a_{13} > c_3$. As $n_1 = a_{12}a_{13} + a_{12}a_{23} + a_{13}a_{32} = a_{13}(a_{12} + a_{32}) + a_{12}a_{23}$ and since by 1) in Lemma 10.24, $a_{12} + a_{32} \geq c_2$, we obtain that $n_1 > c_3 c_2 + a_{12}a_{23} > c_3 c_2$. However, $n_1 = r_{12}(r_{13} + r_{23}) + r_{13}r_{32} = c_2 c_3 - r_{23}r_{32}$ which is smaller than $c_3 c_2$, a contradiction.   □

In view of this, a three embedding dimension numerical semigroup, with pairwise coprime minimal generators, is encoded by the constants r_{ij}'s. We arrange these constants in what we call a 0-matrix.

A *0-matrix* is a matrix A of the form

$$A = \begin{pmatrix} 0 & a_{12} & a_{13} \\ a_{21} & 0 & a_{23} \\ a_{31} & a_{32} & 0 \end{pmatrix}$$

where $a_{12}, a_{13}, a_{21}, a_{23}, a_{31}, a_{32}$ are positive integers such that $\gcd\{A_i, A_j\} = 1$, with

$$A_1 = a_{12}a_{13} + a_{12}a_{23} + a_{13}a_{32},$$
$$A_2 = a_{13}a_{21} + a_{21}a_{23} + a_{23}a_{31},$$
$$A_3 = a_{12}a_{31} + a_{21}a_{32} + a_{31}a_{32}.$$

Theorem 10.26. *If* $A = \begin{pmatrix} 0 & a_{12} & a_{13} \\ a_{21} & 0 & a_{23} \\ a_{31} & a_{32} & 0 \end{pmatrix}$ *is a 0-matrix, then* $\langle A_1, A_2, A_3 \rangle$ *is a numerical semigroup with embedding dimension 3 and whose generators are pairwise relatively prime. Moreover,*

$$(a_{ji} + a_{ki})A_i = a_{ij}A_j + a_{ik}A_k$$

and

$$a_{ji} + a_{ki} = \min\left\{ x \in \mathbb{N} \setminus \{0\} \mid xA_i \in \langle A_j, A_k \rangle \right\}.$$

Conversely, if S is a numerical semigroup with embedding dimension 3 and whose generators are pairwise relatively prime, then there exists a 0-matrix A such that $S = \langle A_1, A_2, A_3 \rangle$.

Proof. As a consequence of Theorem 10.25, we have that $\langle A_1, A_2, A_3 \rangle$ is a numerical semigroup with embedding dimension 3. In view of Lemma 10.24, we have that $(a_{ji} + a_{ki})A_i = a_{ij}A_j + a_{ik}A_k$. From Lemmas 10.19 and 10.23 and Theorem 10.25, we deduce that $a_{ji} + a_{ki} = \min\left\{ x \in \mathbb{N} \setminus \{0\} \mid xA_i \in \langle A_j, A_k \rangle \right\}$.

Assume now that S is a numerical semigroup with minimal system of generators $\{n_1, n_2, n_3\}$ such that $\gcd\{n_i, n_j\} = 1$ for $i \neq j$. Then by Lemma 10.23, we know that $A = \begin{pmatrix} 0 & r_{12} & r_{13} \\ r_{21} & 0 & r_{23} \\ r_{31} & r_{32} & 0 \end{pmatrix}$ is a 0-matrix fulfilling that $S = \langle A_1, A_2, A_3 \rangle$. $\qquad\qquad\square$

Example 10.27 ([81]). $A = \begin{pmatrix} 0 & 1 & 2 \\ 1 & 0 & 1 \\ 4 & 1 & 0 \end{pmatrix}$ is a 0-matrix with $A_1 = 5$, $A_2 = 7$ and $A_3 = 9$.

Hence $\langle 5, 7, 9 \rangle$ is a numerical semigroup with embedding dimension 3. Moreover,

$$5 \times 5 = 1 \times 7 + 2 \times 9,$$
$$2 \times 7 = 1 \times 5 + 1 \times 9,$$
$$3 \times 9 = 4 \times 5 + 1 \times 7.$$

Example 10.28 ([81]). $A = \begin{pmatrix} 0 & 1 & 3 \\ 6 & 0 & 5 \\ 3 & 11 & 0 \end{pmatrix}$ is a 0-matrix with $A_1 = 41$, $A_2 = 63$ and

$A_3 = 102$. Hence $\langle 41, 63, 102 \rangle$ is a numerical semigroup with embedding dimension 3. Moreover,

$$9 \times 41 = 1 \times 63 + 3 \times 102,$$
$$12 \times 63 = 6 \times 41 + 5 \times 102,$$
$$8 \times 102 = 3 \times 41 + 11 \times 63.$$

As a consequence of the above theorem, Lemma 10.20 and Proposition 2.20, we obtain the following explicit description of the pseudo-Frobenius numbers.

Corollary 10.29. *Let* $A = \begin{pmatrix} 0 & a_{12} & a_{13} \\ a_{21} & 0 & a_{23} \\ a_{31} & a_{32} & 0 \end{pmatrix}$ *be a 0-matrix and let* $S = \langle A_1, A_2, A_3 \rangle$.

Then $\mathrm{PF}(S) = \{g_1, g_2\}$, *where*

1) $g_1 = -a_{12}a_{13} - a_{12}a_{23} - a_{12}a_{31} - a_{13}a_{21} - a_{13}a_{32} - a_{21}a_{23} - a_{21}a_{32} - a_{23}a_{31} - a_{31}a_{32} + a_{12}a_{13}a_{21} + a_{12}a_{21}a_{23} + a_{12}a_{13}a_{31} + 2a_{12}a_{23}a_{31} + a_{13}a_{21}a_{32} + a_{21}a_{23}a_{32} + a_{13}a_{31}a_{32} + a_{23}a_{31}a_{32}$,

2) $g_2 = -a_{12}a_{13} - a_{12}a_{23} - a_{12}a_{31} - a_{13}a_{21} - a_{13}a_{32} - a_{21}a_{23} - a_{21}a_{32} - a_{23}a_{31} - a_{31}a_{32} + a_{12}a_{13}a_{21} + a_{12}a_{21}a_{23} + a_{12}a_{13}a_{31} + a_{12}a_{23}a_{31} + 2a_{13}a_{21}a_{32} + a_{21}a_{23}a_{32} + a_{13}a_{31}a_{32} + a_{23}a_{31}a_{32}$.

This can be reformulated as follows in terms of the minimal generators and the c_i's.

Theorem 10.30. *Let S be minimally generated by $\{n_1, n_2, n_3\}$ with $\gcd\{n_i, n_j\} = 1$ for all $i, j \in \{1, 2, 3\}$, with $i \neq j$. Then*

$$\Delta = \sqrt{(c_1 n_1 + c_2 n_2 + c_3 n_3)^2 - 4(c_1 n_1 c_2 n_2 + c_1 n_1 c_3 n_3 + c_2 n_2 c_3 n_3 - n_1 n_2 n_3)}$$

is a positive integer and

$$\mathrm{PF}(S) = \left\{ \frac{1}{2}((c_1 - 2)n_1 + (c_2 - 2)n_2 + (c_3 - 2)n_3 + \Delta), \right.$$
$$\left. \frac{1}{2}((c_1 - 2)n_1 + (c_2 - 2)n_2 + (c_3 - 2)n_3 - \Delta) \right\}.$$

In particular,

$$\mathrm{F}(S) = \frac{1}{2}((c_1 - 2)n_1 + (c_2 - 2)n_2 + (c_3 - 2)n_3 + \Delta).$$

Proof. By using the definitions of the c_i's and r_{ij}'s, together with Lemma 10.23 and Corollary 10.29, one can deduce that $(r_{23}n_3 - r_{32}n_2)^2 = (c_1 n_1 + c_2 n_2 + c_3 n_3)^2 - 4(c_1 n_1 c_2 n_2 + c_1 n_1 c_3 n_3 + c_2 n_2 c_3 n_3 - n_1 n_2 n_3)$. Hence Δ is a positive integer.

From Lemma 10.20 and Proposition 2.20 (with $\mathrm{Ap}(S, n_2)$), we deduce that

$$\mathrm{PF}(S) = \{c_1 n_1 + r_{23}n_3 - (n_1 + n_2 + n_3), c_3 n_3 + r_{21}n_1 - (n_1 + n_2 + n_3)\}.$$

Recall that $c_3 n_3 = r_{31}n_1 + r_{32}n_2$, whence $c_3 n_3 + r_{21}n_1 = (r_{31} + r_{21})n_1 + r_{32}n_2$. By Lemma 10.19, this equals to $c_1 n_2 + r_{32}n_2$. Hence

$$\mathrm{PF}(S) = \{c_1 n_1 + r_{23}n_3 - (n_1 + n_2 + n_3), c_1 n_1 + r_{32}n_2 - (n_1 + n_2 + n_3)\}.$$

We can reformulate this as

$$\mathrm{PF}(S) = \{c_1 n_1 + \max\{r_{23}n_3, r_{32}n_2\} - (n_1 + n_2 + n_3),$$
$$c_1 n_1 + \min\{r_{23}n_3, r_{32}n_2\} - (n_1 + n_2 + n_3)\}$$

By taking into account now that for any $a, b \in \mathbb{N}$, we have that

$$\max\{a,b\} = \frac{(a+b)+\sqrt{(a-b)^2}}{2}, \ \min\{a,b\} = \frac{(a+b)-\sqrt{(a-b)^2}}{2},$$

we obtain that $\mathrm{PF}(S) = \{c_1 n_1 + (r_{23}n_3 + r_{32}n_2 + \Delta)/2 - (n_1 + n_2 + n_3), c_1 n_1 + (r_{23}n_3 + r_{32}n_2 - \Delta)/2 - (n_1 + n_2 + n_3)\} = \{(c_1 n_1 + r_{12}n_2 + r_{13}n_3 + r_{23}n_3 + r_{32}n_2 + \Delta)/2 - (n_1 + n_2 + n_3), (c_1 n_1 + r_{12}n_2 + r_{13}n_3 + r_{23}n_3 + r_{32}n_2 - \Delta)/2 - (n_1 + n_2 + n_3)\} = \{(c_1 n_1 + c_2 n_2 + c_3 n_3 + \Delta)/2 - (n_1 + n_2 + n_3), (c_1 n_1 + c_2 n_2 + c_3 n_3 - \Delta)/2 - (n_1 + n_2 + n_3)\}$. $\square$

Observe that from the proof of this result, we also deduce that $\mathrm{F}(S) = c_1 n_1 + \max\{r_{23}n_3, r_{32}n_2\} - (n_1 + n_2 + n_3)$, which is the well known formula given by Johnson in [42, Theorem 4] (compare also with the expression given in [13]).

By using Selmer's formula for the genus of a numerical semigroup (Proposition 2.12) we obtain the following consequence.

Theorem 10.31. *Let S be minimally generated by $\{n_1, n_2, n_3\}$ with $\gcd\{n_i, n_j\} = 1$ for all $i, j \in \{1, 2, 3\}$, with $i \neq j$.*

$$g(S) = \frac{1}{2}((c_1 - 1)n_1 + (c_2 - 1)n_2 + (c_3 - 1)n_3 - c_1 c_2 c_3 + 1).$$

Proof. From the proof of Lemma 10.23, we know that

$$\mathrm{Ap}(S, n_1) = \{an_2 + bn_3 \mid (a, b) \leq (r_{12} - 1, c_3 - 1) \text{ or } (a, b) \leq (c_2 - 1, r_{13} - 1)\}.$$

By Proposition 2.12,

$$g(S) = \frac{1}{n_1} \sum_{w \in \mathrm{Ap}(S, n_1)} w - \frac{n_1 - 1}{2}.$$

An easy computation yields the desired result. $\square$

By combining this result with Theorem 10.26 we obtain the following.

Corollary 10.32. *Let $A = \begin{pmatrix} 0 & a_{12} & a_{13} \\ a_{21} & 0 & a_{23} \\ a_{31} & a_{32} & 0 \end{pmatrix}$ be a 0-matrix and let $S = \langle \Lambda_1, \Lambda_2, \Lambda_3 \rangle$. Then*

$$g(S) = \frac{1}{2}(1 - a_{12}a_{13} - a_{12}a_{23} - a_{12}a_{31} - a_{13}a_{21}$$

$$- a_{13}a_{32} - a_{21}a_{23} - a_{21}a_{32} - a_{23}a_{31} - a_{31}a_{32}$$

$$+ a_{12}a_{13}a_{21} + a_{12}a_{21}a_{23} + a_{12}a_{13}a_{31} + 2a_{12}a_{23}a_{31}$$

$$+ 2a_{13}a_{21}a_{32} + a_{13}a_{31}a_{32} + a_{21}a_{23}a_{32} + a_{23}a_{31}a_{32}).$$

Exercises

Exercise 10.1. Prove that $\langle 10, 19, 28, 42 \rangle$ is a free numerical semigroup that does not have an Apéry set of unique expression.

Exercise 10.2. Prove that any maximal embedding dimension numerical semigroup has Apéry sets of unique expression.

Exercise 10.3. Let S be a numerical semigroup with minimal system of generators $\{n_1 < n_2 < \cdots < n_e\}$. For every $i \in \{1, \ldots, e\}$, set $c_i = \min\{k \in \mathbb{N} \setminus \{0\} \mid kn_i \in \langle\{n_1, \ldots, n_e\} \setminus \{n_i\}\rangle\}$. We say that S is *simple* if $n_1 = (c_2 - 1) + \cdots + (c_e - 1)$. Prove that

a) $\langle 4, 5, 7 \rangle$ is simple,
b) if S is simple, then $\mathrm{Ap}(S, n_1) = \{0, n_2, \ldots, (c_2 - 1)n_2, \ldots, n_e, \ldots, (c_e - 1)n_e\}$,
c) if S is simple, then it has Apéry sets of unique expression,
d) if S is simple and $n \in S$, then the graph associated to n in S, G_n, is connected if and only if $n = n_i + n_j$ for some $i, j \in \{1, \ldots, e\}$ with $i \neq j$ or $n = c_i n_i$ for some $i \in \{1, \ldots, e\}$.

Exercise 10.4. Let a be a positive integer. Prove that if a is even, then $S = \langle a, a + 1, a + 2 \rangle$ is not pseudo-symmetric. For which values of a is S pseudo-symmetric?

Exercise 10.5. Let S be a symmetric numerical semigroup with embedding dimension three. Prove that there is an arrangement n_1, n_2, n_3 of its minimal generators so that

a) $dn_3 \in \langle n_1, n_2 \rangle$, with $d = \gcd\{n_1, n_2\}$,
b) $\mathrm{F}(S) = \frac{n_1 n_2}{d} - n_1 - n_2 - (d - 1)n_3$.

Exercise 10.6 ([95]). Let S be a proportionally modular symmetric numerical semigroup with embedding dimension three. Prove that there is an arrangement of its minimal generators n_1, n_2, n_3 such that

$$\mathrm{F}(S) = \frac{n_1 n_2 n_3 - n_1 n_2 - n_2 n_2}{n_1 + n_3}.$$

(*Hint:* Prove that $d = \gcd(n_1, n_3) = \frac{n_1 + n_3}{n_2}$ and use the preceding exercise.)

Exercise 10.7 ([96]). Let n_1, n_2 and n_3 be positive integers with $\gcd\{n_1, n_2\} = 1$, and let u be a positive integer such that $un_2 \equiv 1 \bmod n_1$. Set $m = \mathrm{m}(\mathrm{S}(un_2 n_3, n_1 n_2, n_3))$. Prove that mn_3 is the least multiple of n_3 belonging to $\langle n_1, n_2 \rangle$.

Exercise 10.8 ([34]). Prove that if S is an irreducible numerical semigroup with embedding dimension less than or equal to three, then $\mathrm{F}(S) \neq 12$ (*Hint:* Compute the set of irreducible numerical semigroup with Frobenius number 12; there are only two).

Exercise 10.9 ([104]). Let a, b and c be integers greater than one with $\gcd\{a, b\} = \gcd\{a, c\} = \gcd\{b, c\} = 1$. Let n be an integer greater than or equal to three. Prove that the numerical semigroup $\frac{\langle a^n, b^n \rangle}{c}$ is not minimally generated by $\{a^n, c^{n-1}, b^n\}$.

Exercise 10.10 ([81]). Every positive integer is the Frobenius number of a numerical semigroup with at most three generators (compare with Exercise 4.24).

Chapter 10
The structure of a numerical semigroup

Introduction

The aim of this chapter is to study which properties a monoid must fulfill in order
to be isomorphic to a numerical semigroup. Levin shows in [46] that if S is finitely
generated, Archimedean and without idempotents, then S is multiple joined. By
using this as starting point, we will show that a monoid is isomorphic to a numerical
semigroup if and only if it is finitely generated, quasi-Archimedean, torsion free
and with only one idempotent. We will also relate this characterization with other
interesting properties in semigroup theory such as weak cancellativity, being free of
units, and being hereditarily finitely generated.

In [76] the concept of $\mathfrak{N}$-monoid is presented, and generalizing the results given
by Tamura in [103], one can construct all $\mathfrak{N}$-monoids up to isomorphism. In this
chapter we also place numerical semigroups in the scope of $\mathfrak{N}$-monoids.

1 Levin's theorem

An element x in a semigroup S is *Archimedean* if for all $y \in S$ there exists a positive
integer k and $z \in S$ such that $kx = y + z$. We say that S is *Archimedean* if all its
elements are Archimedean.

We say that an element x in a semigroup is *idempotent* if $2x = x$.

A semigroup S is *multiple joined* if for all $x, y \in S$, there exist positive integers p and q such that $px = qy$. Clearly, every multiple joined semigroup is
Archimedean. Our next goal is to prove a result given by Levin, which states that
every Archimedean finitely generated semigroup without idempotents is multiple
joined.

Since in this section we deal with semigroups, we need to translate some of the
concepts already introduced in this book to the more general scope of semigroups.

The following result will be used several times in this chapter. On an Archimedean semigroup without idempotents some nice properties hold.

Lemma 11.1. *Let S be an Archimedean semigroup without idempotents. Then*

1) if $a, b \in S$, then $a \neq a + b$,
2) if $b \in S$, then $\bigcap_{n \in \mathbb{N} \setminus \{0\}} (nb + S)$ is empty.

Proof. Assume to the contrary that $a = a + b$ for some $a, b \in S$. As S is Archimedean, there exist a positive integer k and $c \in S$ such that $kb = a + c$. Hence $(k+1)b = a + c + b = a + c = kb$. It follows that $(k+h)b = kb$ for all $h \in \mathbb{N}$, and in particular $2kb = kb$, contradicting that S has no idempotents.

Assume now that there exists $b \in S$ and $a \in \bigcap_{n \in \mathbb{N} \setminus \{0\}} (nb + S)$. Again, as S is Archimedean, there exist a positive integer k and $c \in S$ such that $kb = a + c$. Since $a \in kb + S$, there exists $d \in S$ such that $a = kb + d$. Hence $a = a + c + d$, contradicting 1). $\square$

As in a monoid, we can define on a semigroup S the binary relation $\leq_S$, but we have to modify it slightly so that it becomes reflexive. For $a, b \in S$, we write $a \leq_S b$ if either $a = b$ or $b = a + c$ for some $c \in S$. Observe that the condition $a = b$ was not needed for monoids, since they have an identity element. Recall that for any numerical semigroup, the relation $\leq_S$ is an order relation. We next see that the same holds for any Archimedean semigroup without idempotents.

Lemma 11.2. *Let S be an Archimedean semigroup without idempotents. Then*

1) $\leq_S$ is an order relation,
2) Minimals$_{\leq_S} S$ is contained in any system of generators of S.

Proof. As pointed out above, reflexivity follows directly from the definition. Antisymmetry follows from Lemma 11.1, and the transitivity holds because S is a semigroup.

Now let $M = \text{Minimals}_{\leq_S}(S)$, and let A be a system of generators of S. Let $m \in M$. Then $m = a_1 + \cdots + a_n$ for some $a_1, \ldots, a_n \in A$. Hence $a_1 \leq_S m$. From the minimality of m, it follows that $a_1 = m$. $\square$

Recall that for a numerical semigroup S, the set Minimals$_{\leq_S}(S \setminus \{0\})$ is the minimal system of generators of S. Note also that $S \setminus \{0\}$ is itself a semigroup and that for all $a, b \in S \setminus \{0\}$, $a \leq_S b$ if and only if $a \leq_{S \setminus \{0\}} b$. Clearly, $S \setminus \{0\}$ is Archimedean and has no idempotents (the only idempotent of S is 0).

We will write $a <_S b$ whenever $a \leq_S b$ and $a \neq b$.

Lemma 11.3. *Let S be an Archimedean finitely generated semigroup without idempotents. Then*

1) there are no sequences $\{a_n\}_{n \in \mathbb{N}} \subseteq S$ such that $a_{i+1} <_S a_i$ for all $i \in \mathbb{N}$,
2) $S = \langle \text{Minimals}_{\leq_S}(S) \rangle$.

Proof. Assume to the contrary that $\{a_n\}_{n\in\mathbb{N}}$ is a sequence of elements in S and that $a_n = a_{n+1} + x_{n+1}$ for some $x_{n+1} \in S$ and for all $n \in \mathbb{N}$. Then for every $k \in \mathbb{N}$, $a_0 = a_k + x_1 + \cdots + x_k$.

Let $\{s_1,\ldots,s_t\}$ be a system of generators of S. Then $x_1 + \cdots + x_k = p_{k_1}s_1 + \cdots + p_{k_t}s_t$ for some $p_{k_1},\ldots,p_{k_t} \in \mathbb{N}$. For all $j \in \{1,\ldots,t\}$, let $R_j = \lim_{k\to\infty} p_{k_j}$ and assume that there exists $R = \max\{R_1,\ldots,R_t\} < \infty$. Observe that the sequence $\{p_{k_j}\}_{k\geq 1}$ is nondecreasing. Hence the finiteness of R implies that the set $\{x_1 + \cdots + x_k\}_{k\geq 1}$ is finite. Thus there exist positive integers k and h such that $x_1 + \cdots + x_k = x_1 + \cdots + x_k + x_{k+1} + \cdots + x_{k+h}$, contradicting Lemma 11.1. Hence there exists j such that $R_j = \infty$. It follows that $a_0 = a_j + y_1 = 2a_j + y_2 = \cdots = na_j + y_n = \cdots$ for some $\{y_1,y_2,\ldots,y_n,\ldots\} \subseteq S$. This implies that $a_0 \in \bigcap_{n\geq 1}(na_j + S)$, contradicting once more Lemma 11.1.

Condition 2) follows as in the second part of Proposition 8.5. The proof is left to the reader. $\qquad\square$

Let S be a semigroup. A *congruence* on S is an equivalence binary relation compatible with the binary operation on S. As with monoids, if S is a semigroup and σ is a congruence on S, then the *quotient set* S/σ is a semigroup, called the *quotient semigroup* of S by σ.

Given a semigroup S and $b \in S$ define the binary relation σ_b on S as follows:

$$x\sigma_b y \text{ if } x + nb = y + mb \text{ for some } n,m \in \mathbb{N}\setminus\{0\}.$$

In order to simplify notation we will write $[a]$ for the σ_b-class of a in S. This binary relation has some nice properties.

Lemma 11.4. *Let S be a semigroup and let b be an element of S. Then*

1) σ_b is a congruence on S,
2) $[b]$ is the identity element of S/σ_b,
3) $[b]$ is a subsemigroup of S,
4) if b is an Archimedean element of S, then $\frac{S}{\sigma_b}$ is a group.

Proof. For every $x \in S$, $x + b = x + b$, and thus σ_b is reflexive. If $x + nb = y + mb$, then $y + mb = x + nb$, whence σ_b is symmetric. Transitivity follows from the fact that if $x + nb = y + mb$ and $y + kb = z + lb$, then $x + (n+k)b = z + (m+l)b$. Finally, if $x + nb = y + mb$, then $(x+z) + nb = (y+z) + mb$ for all $x,y,z \in S$ and $n,m \in \mathbb{N}\setminus\{0\}$.

As $x + b + b = x + 2b$, we have that $[x] + [b] = [x+b] = [x]$, which means that $[b]$ is the identity element in the quotient semigroup $\frac{S}{\sigma_b}$.

If $x,y \in [b]$, then $x + nb = mb$ and $y + kb = lb$ for some positive integers n, m, k and l. Hence $x + y + (n+k)b = (m+l)b$, and thus $x+y \in [b]$. This proves that $[b]$ is a subsemigroup of S.

Assume now that b is an Archimedean element of S. Let $x \in S$. Then there exists $y \in S$ and a positive integer k such that $kb = x+y$. Hence $[x] + [y] = [kb] = [b]$, which proves that $[y]$ is the inverse of $[x]$. $\qquad\square$

We can still sharpen these properties for Archimedean finitely generated semigroups without idempotents.

Lemma 11.5. *Let S be an Archimedean finitely generated semigroup without idempotents, and let $b \in S$. Set $I_b = S \setminus (b+S)$. Then*

1) if $[a] \in \frac{S}{\sigma_b}$, then $[a] \cap I_b$ is not empty,
2) the set I_b has finitely many elements,
3) $\frac{S}{\sigma_b}$ is a finite group,
4) if $x \in b+S$, there exists $p_x \in I_b$ and a positive integer k_x such that $x = p_x + k_x b$
   (and thus b and p_x are in the same σ_b-class).

Proof. Let $a \in S$. If $a \in b+S$, then $a = b+x_1$ for some $x_1 \in S$. In particular, $[a] = [x_1]$. If $x_1 \notin b+S$, then we are done. Otherwise, $x_1 = b+x_2$, for some $x_2 \in S$. Once we have defined x_i, if $x_i \in b+S$, then set $x_{i+1} = x_i$, otherwise $x_i = b+x_{i+1}$ for some $x_i \in S$. By Lemma 11.1, $x_{i+1} \neq x_{i+1}+b$, and thus $x_i = x_{i+1}$ if and only if $x_i \in b+S$. In this way a sequence $a \geq_S x_1 \geq_S x_2 \geq_S \cdots$ of elements in $[a]$, which in view of Lemma 11.3 must be stationary. Thus there exist i such that $x_i = x_{i+1}$, which as we have pointed out above means that $x_i \in (b+S) \cap [a]$. This proves 1). Note also that in this setting $a = ib+x_i$, which also proves 4).

Assume now that I_b has infinitely many elements and let $\{a_n\}_{n\in\mathbb{N}} \subseteq I_b$ with $a_i \neq a_j$ for $i \neq j$. By Lemma 11.3, $S = \langle \text{Minimals}_{\leq_S}(S) \rangle$, and by Lemma 11.2, we know that $\text{Minimals}_{\leq_S}(S)$ must have finitely many elements because S is finitely generated. Let $\text{Minimals}_{\leq_S}(S) = \{a_1, \ldots, a_t\}$. There exists for all $i \in \mathbb{N}$, $\lambda_{i_1}, \ldots, \lambda_{i_t} \in \mathbb{N}$ such that $x_i = \lambda_{i_1} a_1 + \cdots + \lambda_{i_t} a_t$. As the sequence $\{x_n\}_{n\in\mathbb{N}}$ is infinite, there exists $i \in \{1, \ldots, t\}$ such that $\{\lambda_{k_i}\}_{k\in\mathbb{N}}$ is not bounded. Since S is Archimedean, there exists a positive integer l and $x \in S$ such that $la_i = b+x$. Choose k big enough such that $\lambda_{k_i} > l$. Then $x_k = \lambda_{k_1} a_1 + \cdots + \lambda_{k_i} a_i + \cdots + \lambda_{k_t} a_t = b+y$ for some $y \in S$, contradicting that $x_k \in I_b$. In this way we have shown that Condition 2) is true.

From 1) and 2) we have that S/σ_b is finite. From Lemma 11.4 we deduce that it is a finite group. □

These were the tools needed to prove Levin's theorem.

Theorem 11.6 ([46]). *Let S be an Archimedean finitely generated semigroup without idempotents. Then S is multiple joined.*

Proof. Let $x, y, b \in S$. From Lemma 11.5, the semigroup S/σ_b is a finite group and by Lemma 11.4, $[b]$ is its identity element. Hence there exist positive integers such that $n[x] = [b] = m[y]$. This in particular implies that $\{nx, my\} \subseteq [b]$. Now let $\{p_1, \ldots, p_r\} = I_b \cap [b]$ (this set is finite by Lemma 11.5). Then there exist positive integers $n_1, \ldots, n_r, k_1, \ldots, k_r$ such that $p_1 + n_1 b = k_1 b, p_2 + n_2 b = k_2 b, \ldots, p_r + n_r b = k_r b$. Lemma 11.1 forces n_i to be less than k_i for all $i \in \{1, \ldots, r\}$. As S does not have idempotents, by Lemma 11.1, we deduce that the elements in the sequence $\{knx\}_{k\geq 1}$ are all different. Since I_b is finite, there exists $u \in \mathbb{N} \setminus \{0\}$ such that $unx \notin I_b$, or equivalently, $unx \in b+S$. By Lemma 11.5, there exists $i \in \{1, \ldots, r\}$ and a positive integer t such that $unx = p_i + tb$. Note that if $t \leq n_i$, then $kunx + sb = kp_i + kn_i b = p_i + (k-1)(p_i + n_i b) + n_i b = p_i + ((k-1)k_i + n_i)b$, and now $(k-1)k_i + n_i > n_i$ for k positive. Thus we can also assume that $t > n_i$. Hence $unx = p_i + n_i b + (t-n_i)b = k_i b + (t-n_i)b = \alpha b$ for some $\alpha \in \mathbb{N} \setminus \{0\}$. Analogously, we can find $v, \beta \in \mathbb{N} \setminus \{0\}$ such that $vmy = \beta b$. Thus $\beta unx = \alpha vmy$. □

2 Structure theorem

Our aim in this section is to prove Theorem 11.18, which characterizes those monoids that are isomorphic to a numerical semigroup. On the way to this result some interesting properties of numerical semigroups will show up.

Observe that if on a monoid M, the identity element $0 \in M$ is Archimedean, then for every $x \in M$ there exist a positive integer k and $y \in M$ such that $k0 = x + y$. Hence $0 = x + y$ and thus x has inverse y. This in particular implies that if a monoid is Archimedean, then it is a group. Clearly, the converse is also true, because for a and b in a group G, $1a = b + (-b + a)$ (recall that we are omitting the adjective commutative).

Lemma 11.7. *A monoid is Archimedean if and only if it is a group.*

Recall that if S is a numerical semigroup, then $S \setminus \{0\}$ is Archimedean, and 0 is trivially not Archimedean. These are the monoids we are interested in, since Archimedean monoids are groups as observed above. Let A be a monoid. We say that A is *quasi-Archimedean* if 0 is not an Archimedean and the rest of the elements in A are Archimedean. Hence every numerical semigroup is quasi-Archimedean. Note that $\{0\}$ is not quasi-Archimedean, since 0 in $\{0\}$ is Archimedean. Thus any quasi-Archimedean monoid is nontrivial.

An element a in a monoid A is a *unit* if there exist $b \in A$ such that $a + b = 0$. A monoid is *free of units* if its only unit is the identity element. It is easy to prove the following result.

Lemma 11.8. *Let A be a monoid free of units. Then $A \setminus \{0\}$ is a semigroup. Moreover, if A is finitely generated as a monoid, then so is $A \setminus \{0\}$ as a semigroup.*

Quasi-Archimedean monoids are free of units as we see in the following result (as occurs with numerical semigroups).

Lemma 11.9. *Every quasi-Archimedean monoid is free of units.*

Proof. Let A be a quasi-Archimedean monoid. Assume that there exist $a, b \in A \setminus \{0\}$ such that $a + b = 0$. Let c be an element of A. As $a \neq 0$, it is Archimedean, and there exist a positive integer k and $d \in A$ such that $ka = c + d$. Hence $k0 = ka + kb = c + d + kb$, which proves that 0 is Archimedean, contradicting the hypothesis. $\square$

As a consequence of Lemma 11.1, we obtain the following result.

Lemma 11.10. *Let S be an Archimedean semigroup without idempotents. Let $a \in S$ and $k_1, k_2 \in \mathbb{N} \setminus \{0\}$ be such that $k_1 a = k_2 a$. Then $k_1 = k_2$.*

A semigroup S is *torsion free* if for any positive integer k and $a, b \in S$ such that $ka = kb$, then $a = b$. With this we close the first list of properties that characterize those monoids isomorphic to a numerical semigroup.

Proposition 11.11. *Let A be a monoid. The following conditions are equivalent.*

1) A is isomorphic to a numerical semigroup.
2) A is finitely generated, quasi-Archimedean, torsion free and with only one idempotent.

Proof. *1) implies 2).* We have already mentioned that a numerical semigroup is quasi-Archimedean. Clearly, it is also torsion free, and its only idempotent is 0. By Theorem 2.7 it is finitely generated.

2) implies 1). Let $\{m_1,\ldots,m_e\}$ be a system of generators of $A \setminus \{0\}$ as a semigroup (Lemma 11.8), and thus it is a system of generators of A with $m_i \neq 0$ for all $i \in \{1,\ldots,e\}$. Observe that $A \setminus \{0\}$ is Archimedean and has no idempotents. By Levin's theorem (Theorem 11.6) we know that $A \setminus \{0\}$ is multiple joined. Hence there exist positive integers $a_2,\ldots,a_e,b_2,\ldots,b_e$ such that $a_2m_1 = b_2m_2,\ldots,a_em_1 = b_em_e$. Let $M = b_2 \cdot b_3 \cdots b_e$ and let $S = \langle M, \frac{M}{b_2}a_2, \frac{M}{b_3}a_3,\ldots,\frac{M}{b_e}a_e\rangle \subseteq \mathbb{N}$. As every nontrivial submonoid of $\mathbb{N}$ is isomorphic to a numerical semigroup (Proposition 2.2), we only have to show that A is isomorphic to S. Define

$$f : A \to S, \; f(\lambda_1 m_1 + \cdots + \lambda_e m_e) = \lambda_1 M + \lambda_2 \frac{M}{b_2}a_2 + \cdots + \lambda_e \frac{M}{b_e}a_e.$$

- Let us see that f is a map, that is, if $\lambda_1 m_1 + \cdots + \lambda_e m_e = \mu_1 m_1 + \cdots + \mu_e m_e$, then $\lambda_1 M + \lambda_2 \frac{M}{b_2}a_2 + \cdots + \lambda_e \frac{M}{b_e}a_e = \mu_1 M + \mu_2 \frac{M}{b_2}a_2 + \cdots + \mu_e \frac{M}{b_e}a_e$. Note that $(\lambda_1 M + \lambda_2 \frac{M}{b_2}a_2 + \cdots + \lambda_e \frac{M}{b_e}a_e)m_1 = \lambda_1 Mm_1 + \lambda_2 \frac{M}{b_2}a_2m_1 + \cdots + \lambda_e \frac{M}{b_e}a_em_1 = M(\lambda_1 m_1 + \cdots + \lambda_e m_e)$. Analogously, one shows that $(\mu_1 M + \mu_2 \frac{M}{b_2}a_2 + \cdots + \mu_e \frac{M}{b_e}a_e)m_1 = M(\mu_1 m_1 + \cdots + \mu_e m_e)$. By Lemma 11.10, as $(\lambda_1 M + \lambda_2 \frac{M}{b_2}a_2 + \cdots + \lambda_e \frac{M}{b_e}a_e)m_1 = (\mu_1 M + \mu_2 \frac{M}{b_2}a_2 + \cdots + \mu_e \frac{M}{b_e}a_e)m_1$, we deduce that $\lambda_1 M + \lambda_2 \frac{M}{b_2}a_2 + \cdots + \lambda_e \frac{M}{b_e}a_e = \mu_1 M + \mu_2 \frac{M}{b_2}a_2 + \cdots + \mu_e \frac{M}{b_e}a_e$.
- Let us prove that it is injective. If $f(\lambda_1 m_1 + \cdots + \lambda_e m_e) = f(\mu_1 m_1 + \cdots + \mu_e m_e)$, then $f(\lambda_1 m_1 + \cdots + \lambda_e m_e)m_1 = f(\mu_1 m_1 + \cdots + \mu_e m_e)m_1$. Arguing as in the preceding paragraph, we get that $M(\lambda_1 m_1 + \cdots + \lambda_e m_e) = M(\mu_1 m_1 + \cdots + \mu_e m_e)$. As A is torsion free, we get that $\lambda_1 m_1 + \cdots + \lambda_e m_e = \mu_1 m_1 + \cdots + \mu_e m_e$.
- The map f is surjective by definition and it is clearly a monoid morphism.

Hence f is a monoid isomorphism. □

Recall that we used that every numerical semigroup is cancellative in order to show that it was finitely presented. We will try to include this property in the list of conditions we must impose to a monoid so that it is isomorphic to a numerical semigroup.

The identity element is always an idempotent of any monoid. If the monoid is cancellative, this is the only idempotent.

Lemma 11.12. *The only idempotent of a cancellative monoid is its identity element.*

Proof. Let A be a cancellative monoid. Let x be an idempotent of A. Hence $2x = x$, and thus $x + x + 0 = x + 0$. By cancelling in both sides x, we get that $x = 0$. □

As a consequence of this result and Proposition 11.11, we obtain an alternative characterization of monoids isomorphic to numerical semigroups.

Proposition 11.13. *A monoid is isomorphic to a numerical semigroup if and only if it is finitely generated, quasi-Archimedean, cancellative and torsion free.*

These conditions can be slightly modified and still get the same result. In particular, we are going to perturb the cancellative property, and substitute it by a lighter condition.

We say that a monoid A is *weakly cancellative* if given $x, y \in A$ such that $x + a = y + a$ for all $a \in A \setminus \{0\}$, one gets that $x = y$. Every cancellative monoid is weakly cancellative. The converse does not hold in general, though under certain circumstances it does.

Lemma 11.14. *Let A be a quasi-Archimedean weakly cancellative finitely generated monoid. Then A is cancellative.*

Proof. Let $\{m_1, \ldots, m_e\}$ be a system of generators of A such that $s_i \neq 0$ for all $i \in \{1, \ldots, e\}$ (if $s_i = 0$ for some i, then it is not needed in the system of generators and thus we can remove it). Assume that there exist $s, t, x \in A$ such that $s + x = t + x$ and $s \neq t$. As A is quasi-Archimedean, for all $i \in \{1, \ldots, e\}$ there exist a positive integer k_i and $y_i \in A$ such that $k_i m_i = x + y_i$. Hence, $s + (a_1 m_1 + \cdots + a_e m_e) \neq t + (a_1 m_1 + \cdots + a_e m_e)$, forces $(a_1, \ldots, a_e) \leq (k_1, \ldots, k_e)$. Thus the set

$$M = \text{Maximals}_{\leq} \{(a_1, \ldots, a_e) \in \mathbb{N}^e \mid$$
$$s + (a_1 m_1 + \cdots + a_e m_e) \neq t + (a_1 m_1 + \cdots + a_e m_e)\}$$

is finite. If $(d_1, \ldots, d_e) \in M$, then $s + (d_1 m_1 + \cdots + d_e m_e) \neq t + (d_1 m_1 + \cdots + d_e m_e)$ and $s + (d_1 m_1 + \cdots + d_e m_e) + m_i = t + (d_1 m_1 + \cdots + d_e m_e) + m_i$ for all $i \in \{1, \ldots, e\}$. This implies that $s + (d_1 m_1 + \cdots + d_e m_e) + a = t + (d_1 m_1 + \cdots + d_e m_e) + a$ for all $a \in A \setminus \{0\}$. By using now that A is weakly cancellative, we deduce that $s + (d_1 m_1 + \cdots + d_e m_e) = t + (d_1 m_1 + \cdots + d_e m_e)$, contradicting that $(d_1, \ldots, d_e) \in M$. □

As a consequence of this result and Proposition 11.13, we deduce a new characterization for the monoids that are isomorphic to a numerical semigroup.

Proposition 11.15. *A monoid is isomorphic to a numerical semigroup if and only if it is finitely generated, quasi-Archimedean, weakly cancellative and torsion free.*

We still can modify the conditions imposed in the preceding propositions. Observe that any submonoid S of a numerical semigroup is a submonoid of $\mathbb{N}$ and thus it is finitely generated (Corollary 2.8). This property does not hold for monoids in general (see Exercise 2.15). A monoid fulfilling that any of its submonoids is finitely generated is a *hereditarily finitely generated* monoid. Thus any numerical semigroup is hereditarily finitely generated.

Lemma 11.16. *Let A be a nontrivial monoid that is hereditarily finitely generated, cancellative, torsion free and free of units. Then A is quasi-Archimedean.*

Proof. As A is free of units, 0 is not Archimedean (see the paragraph preceding Lemma 11.7). Let $x \in A \setminus \{0\}$ (we are assuming that $A \neq \{0\}$) and let $y \in A$. The set $H = \{ax + by \mid a, b \in \mathbb{N} \setminus \{0\}\} \cup \{0\}$ is a submonoid of A, and thus it must be finitely generated.

First we show that all the elements in $X = \{ax + y \mid a \in \mathbb{N} \setminus \{0\}\}$ are different. Assume to the contrary that there exist $a, b \in \mathbb{N} \setminus \{0\}$, with for instance $a < b$, such that $ax + y = bx + y$. As A is cancellative, this yields $(b - a)x = 0 = (b - a)0$, and $x \neq 0$, contradicting that A is torsion free. In particular this implies that $\{ax + y \mid a \in \mathbb{N} \setminus \{0\}\}$ must be infinite. By using that H is finitely generated and that all the elements in X are different, we deduce that there exist an equality of the form $ax + y = cx + dy$ with $d > 1$. By using again that A is cancellative, we get that $ax = cx + (d - 1)y$. If $c \geq a$, we obtain that $(c - a)x + (d - 1)y = 0$, contradicting either that A is free of units (if $c - a > 0$) or that it is torsion free (if $c = a$; note that $d > 1$). Hence $c < a$ and consequently $(a - c)x = (d - 2)y + y$. This proves that x is Archimedean. □

This, together with Proposition 11.13 yields another characterization of those monoids that are isomorphic to a numerical semigroup.

Proposition 11.17. *A nontrivial monoid is isomorphic to a numerical semigroup if and only if it is hereditarily finitely generated, cancellative, torsion free and free of units.*

By gathering all these characterizations, we obtain the following.

Theorem 11.18. *Let A be a monoid. The following conditions are equivalent.*

1) A is isomorphic to a numerical semigroup.
2) A is finitely generated, quasi-Archimedean, torsion free and with only one idempotent.
3) A is finitely generated, quasi-Archimedean, cancellative and torsion free.
4) A is finitely generated, quasi-Archimedean, weakly cancellative and torsion free.
5) A is hereditarily finitely generated, cancellative, torsion free and free of units.

3 $\mathfrak{N}$-monoids

We now relate numerical semigroups with a concept that has been well studied in the semigroup literature.

Let S be a cancellative monoid. Define on $S \times S$ the following binary relation:

$$(a, b) \sim (c, d) \text{ if } a + d = b + c.$$

Note the similarity of this definition with that of the congruence used on a domain to define its quotient field.

Proposition 11.19. *Let S be a cancellative monoid and let $\sim$ be defined as above.*

1) The relation $\sim$ is a congruence.
2) $(S \times S)/\sim$ is a group.
3) S is isomorphic to a submonoid of $(S \times S)/\sim$.

Proof. Condition 1) is straightforward to prove. Clearly $[(0,0)]$ is the identity element of the monoid $(S \times S)/\sim$. For $[(a,b)] \in (S \times S)/\sim$, $[(a,b)] + [(b,a)] = [(a+b,a+b)] = [(0,0)]$. Hence every element has an inverse. This proves 2).

Define $i : S \rightarrow (S \times S)/\sim$, $i(a) = [(a,0)]$. It is easy to see that i is a monoid homomorphism. If $i(a) = i(b)$, then $[(a,0)] = [(b,0)]$, or equivalently, $(a,0) \sim (b,0)$. By definition this means that $a+0 = b+0$. Thus $a = b$, and i is injective. This shows 3) because S and $\mathrm{im}(i)$ are isomorphic monoids. $\qquad\qquad\square$

We will denote the group $(S \times S)/\sim$ by $Q(S)$ and will call it the *quotient group* of S.

Lemma 11.20. *If two cancellative monoids are isomorphic, then so are their quotient groups.*

Proof. Let $f : S \rightarrow T$ be a monoid isomorphism between the monoids S and T. Let $Q(S)$ and $Q(T)$ be the quotient groups of S and T, respectively. Recall that the map $i : T \rightarrow Q(T)$, $t \mapsto [(t,0)]$ embeds T as a submonoid of $Q(T)$. Define $g : Q(S) \rightarrow Q(T)$ by $g([(s_1,s_2)]) = i(f(s_1)) - i(f(s_2))$. Note that $g([(s_1,s_2)]) = [(f(s_1),f(s_2))]$. As $[(s_1,s_2)] = [(r_1,r_2)]$ if and only if $s_1 + r_2 = s_2 + r_1$, and as f is a monoid isomorphism, this holds if and only if $f(s_1) + f(r_2) = f(s_2) + f(r_1)$, or equivalently, $[(f(s_1),f(s_2))] = [(f(r_1),f(r_2))]$. This proves that g is a map and it is injective. Clearly, g is surjective because f is surjective. Finally $g([(s_1,s_2)] + [(r_1,r_2)]) = g([(s_1+r_1,s_2+r_2)]) = [(f(s_1+r_1),f(s_2+r_2))]$. By using that f is a monoid homomorphism, this equals $[(f(s_1)+f(r_1),f(s_2)+f(r_2))] = g([(s_1,s_2)]) + g([(r_1,r_2)])$. $\qquad\square$

As a consequence of Proposition 11.19 we obtain a nice characterization of cancellative monoids.

Corollary 11.21. *A monoid is cancellative if and only if it is isomorphic to a submonoid of a group.*

An $\mathfrak{N}$-*monoid* is a cancellative monoid that is not a group and with at least an Archimedean element. Next, we see how we can construct the set of all $\mathfrak{N}$-monoids. This construction relies on the choice of a group and a map I fulfilling certain conditions.

In this section and unless otherwise stated, G is a group and $I : G \times G \rightarrow \mathbb{N}$ is a map such that

P.1) $I(g_1,g_2) = I(g_2,g_1)$ for all $g_1,g_2 \in G$,
P.2) $I(g_1,g_2) + I(g_1 + g_2,g_3) = I(g_2,g_3) + I(g_1,g_2 + g_3)$ for all $g_1,g_2,g_3 \in G$,
P.3) $I(0,0) = 0$.

These conditions will make sense soon, because this map behaves like a carry operation. Next we see that $I(0,g) = 0$ for all $g \in G$.

Lemma 11.22. $I(0,g) = 0$ *for all* $g \in G$.

Proof. Use P.2) with $g_1 = g_2 = 0$ and $g_3 = g$. Then $I(0,0) + I(0,g) = I(0,g) + I(0,g)$. Hence $I(0,g) = I(0,0)$, which is 0 by P.3).   □

If M is a monoid contained in a group H (and thus it is cancellative), then the set $\{a-b \mid a,b \in M\}$ is a subgroup of H. The map

$$\frac{M \times M}{\sim} \to \{a-b \mid a,b \in M\}, [(a,b)] \mapsto a-b$$

is a group isomorphism (note that $a - b = c - d$ if and only if $a + d = b + c$, which means that $[(a,b)] = [(c,d)]$). So we will sometimes identify $Q(M)$ with this set.

Proposition 11.23. *The set* $\mathbb{Z} \times G$ *is a group with the binary operation*

$$(z_1,g_1) +_I (z_2,g_2) = (z_1 + z_2 + I(g_1,g_2), g_1 + g_2).$$

Moreover, $(\mathbb{N} \times G, +_I)$ *is a monoid and* $(\mathbb{Z} \times G, +_I)$ *is its quotient group.*

Proof. By using P.1), P.2) and P.3), it is easy to see that $+_I$ is associative and commutative, and also that $(0,0)$ is the identity element in view of Lemma 11.22. Moreover, $(z,g) +_I (-z - I(g,-g), -g) = (0,0)$, and consequently $(\mathbb{Z} \times G, +_I)$ is a group.

Clearly $\mathbb{N} \times G$ is closed under $+_I$ and $(0,0) \in \mathbb{N} \times G$. Thus it is a submonoid of $(\mathbb{Z} \times G, +_I)$.

Now take $(z,g) \in \mathbb{Z} \times G$ and let $a,b \in \mathbb{N}$ be such that $a - b = z$. Note that $(a,0) +_I (-a,0) = (0,0)$. Hence $(b,g) -_I (a,0) = (b,g) +_I (-a,0) = (b - a + I(0,g),g)$ which by Lemma 11.22 equals $(b - a,g) = (z,g)$. This implies that $(\mathbb{Z} \times G, +_I)$ is the quotient group of $(\mathbb{N} \times G, +_I)$.   □

The following result is similar to Lemma 11.9.

Lemma 11.24. *Let S be a monoid. Assume that S has a unit that is Archimedean. Then S is a group.*

Proof. Let u be a unit of S that is Archimedean. Assume that $v \in S$ is such that $u + v = 0$. Let $s \in S$. As u is Archimedean, there exist a positive integer k and $t \in S$ such that $ku = s + t$. Hence $0 = k0 = k(u + v) = ku + kv = s + t + kv$, which implies that $t + kv$ is the inverse of s.   □

If S is a cancellative monoid, then by Corollary 11.21, S is a submonoid of some group G. Let $a \in S$. Then it makes sense to write $-a$ when we refer to the inverse of a in G. Note also that if $a + b = c$ for some $a,b,c \in S$, then we can write $a = c - b \in S$. In particular, if a is a unit of S, then we can denote by $-a$ its inverse.

Given $(a,g) \in \mathbb{Z} \times G$ and $n \in \mathbb{N}$, we will write for sake of simplicity $n(a,g)$ for $(a,g) +_I \cdots +_I (a,g)$ (n times). If n is a negative integer, $n(a,g)$ represents the inverse of $(-n)(a,g)$ in $(\mathbb{Z} \times G, +_I)$.

Theorem 11.25. *Let G be a group and let $I : G \times G \to \mathbb{N}$ be a map fulfilling conditions P.1), P.2) and P.3). Then $(\mathbb{N} \times G, +_I)$ is an $\mathfrak{N}$-monoid. Moreover, every $\mathfrak{N}$-monoid is of this form.*

Proof. As $(\mathbb{N} \times G, +_I)$ is a submonoid of the group $(\mathbb{Z} \times G, +_I)$ (Proposition 11.23), we have that $(\mathbb{N} \times G, +_I)$ is cancellative. Let us see that it is not a group. And for this, we show that $(1,0)$ has no inverse. If $(1,0) +_I (n,g) = (0,0)$ for some $(n,g) \in \mathbb{N} \times G$, then $(n + 1 + I(0,g), g) = (n + 1, g) = (0,0)$, which is impossible. Moreover, let us see that $(1,0)$ is an Archimedean element. By induction, it easily follows that $k(1,0) = (k,0)$, for any nonnegative integer k. Let $(n,g) \in \mathbb{N} \times G$. Then $(n,g) +_I (0,-g) = (n + I(g,-g), 0) = (n + I(g,-g))(1,0)$. This proves that $(\mathbb{N} \times G, +_I)$ is an $\mathfrak{N}$-monoid.

Let S be an $\mathfrak{N}$-monoid, and let b be an Archimedean element of S. Define σ_b as in page 157. By Lemma 11.4, S/σ_b is a group. This group will play the role of G in our representation of S as $(\mathbb{N} \times G, +_I)$.

Let $x \in S$. As b is an Archimedean element, there exist $k \in \mathbb{N} \setminus \{0\}$ and $y \in S$ such that $kb = x + y$. Assume that there exist $l > k$ such that $x - lb \in S$. Then $(kb - x) + (x - lb) = (k - l)b \in S$. And as $l > k$, $(l - k)b$ also belongs to S. Thus $b + (k - l)b + (l - k - 1)b = 0$, which implies that b is a unit of S. By Lemma 11.24, S is a group, contradicting that S is an $\mathfrak{N}$-monoid. Thus implies that the set $\{n \in \mathbb{N} \mid x - nb \in S\}$ is finite. Hence it has a maximum, which we will denote by k_x.

We now prove that if $x\sigma_b y$, then $x + k_y b = y + k_x b$. If $x\sigma_b y$, then by Lemma 11.4, $(x + k_y b)\sigma_b(y + k_x b)$. By definition, there exist positive integers k and l such that $x + k_y b + kb = y + k_x b + lb$. If $l > k$, then $x - (k_x + l - k)b = y - k_y b \in S$, contradicting the maximality of k_x. Analogously k cannot be greater than l, and consequently $k = l$. By using that S is cancellative, we obtain that $x + k_y b = y + k_x b$.

In particular, the above paragraph implies that $\theta : \frac{S}{\sigma_b} \to S$, defined by $\theta([x]) = x - k_x b$, is a map. Note that $\operatorname{im}(\theta) = \{s \in S \mid s - b \notin S\}$ (compare this with the definition of Apéry set of an element in a numerical semigroup).

We see next that for any $s \in S$, there exists a unique $(k,x) \in \mathbb{N} \times \operatorname{im}(\theta)$ such that $s = kb + x$ (compare with Lemma 2.6). Clearly, $s = k_s b + (s - k_s b)$, and this proves the existence. For the uniqueness, assume that $kb + x = lb + y$ for some $k, l \in \mathbb{N}$ and $x, y \in \operatorname{im}(\theta)$. If $l > k$, then $x - (l - k)b \in S$, contradicting that $x \in \operatorname{im}(\theta)$. A similar contradiction occurs if $l < k$, whence $k = l$ and as S is cancellative $x = y$.

Note that if $x, y \in S$, then $(x + y) - (k_x + k_y)b \in S$, which implies that $k_{x+y} \geq k_x + k_y$. Also, if $[x] = [x']$ and $[y] = [y']$ in S/σ_b, then as shown above, $x + k_{x'} b = x' + k_x b$, $y + k_{y'} b = y' + k_y b$ and $(x + y) + k_{x'+y'} b = (x' + y') + k_{x+y} b$. It follows that $k_{x+y} - k_x - k_y = k_{x'+y'} - k_{x'} - k_{y'}$. Hence we can define the map $I : \frac{S}{\sigma_b} \times \frac{S}{\sigma_b} \to \mathbb{N}$ as $I([x],[y]) = k_{x+y} - k_x - k_y$. Let us show that I fulfills Conditions P.1), P.2) and P.3). The first condition follows trivially, and the second after an easy computation. Observe that $k_0 = 0$, since otherwise b would be a unit which we already know is impossible. From this remark we easily obtain P.3).

Finally let us prove that S is isomorphic to $(\mathbb{N} \times \frac{S}{\sigma_b}, +_I)$. The isomorphism is defined by $s \mapsto (k, [x])$, with (k,x) the unique element in $\mathbb{N} \times \operatorname{im}(s)$ such that $s =$

$x + kb$. As $[x] = [s]$ and $k = k_s$, this map can also be written as $s \mapsto (k_s, [s])$. By using this, it is easy to show that this map is a monoid isomorphism. ☐

From the definition, it follows that every numerical semigroup is an $\mathfrak{N}$-monoid. Now that we know how to construct all $\mathfrak{N}$-monoids, we focus on those that are isomorphic to a numerical semigroup. That is, we see which conditions G and I must fulfill so that $(\mathbb{N} \times G, +_I)$ is isomorphic to a numerical semigroup. First we see that G must be cyclic and with finitely many elements.

Proposition 11.26. *Let G be a group and let I be a map fulfilling P.1), P.2) and P.3). If $(\mathbb{N} \times G, +_I)$ is isomorphic to a numerical semigroup, then G is a finite cyclic group.*

Proof. Note that the quotient group of a numerical semigroup is (isomorphic to) $\mathbb{Z}$. Hence if $(\mathbb{N} \times G, +_I)$ is isomorphic to a numerical semigroup, in view of Lemma 11.20 and Proposition 11.23, $(\mathbb{Z} \times G, +_I)$ is isomorphic to $\mathbb{Z}$. As $\mathbb{Z}$ is cyclic, then so is $(\mathbb{Z} \times G, +_I)$. Let (a, α) be a generator of $(\mathbb{Z} \times G, +_I)$. Then for every $g \in G$, there exists $k \in \mathbb{Z}$ such that $(0, g) = k(a, \alpha)$. Hence $g = k\alpha$. This proves that G is generated by α and thus it is cyclic.

By Theorem 11.18, we know that $(\mathbb{N} \times G, +_I)$ is hereditarily finitely generated. Let $H = \{(a, g) \mid a \in \mathbb{N} \setminus \{0\}, g \in G\} \cup \{(0, 0)\}$. Then H is a submonoid of $(\mathbb{N} \times G, +_I)$ and every system of generators of H must contain the set $\{(1, g) \mid g \in G\}$. This forces G to be finite. ☐

Lemma 11.27. *Let G be a finite group. Then the monoid $(\mathbb{N} \times G, +_I)$ is finitely generated.*

Proof. Note that $(a, g) = a(1, 0) +_I (0, g)$. Hence the set $\{(0, g) \mid g \in G\} \cup \{(1, 0)\}$ is a system of generators of $(\mathbb{N} \times G, +_I)$. ☐

Next we see the condition I must fulfill so that $(\mathbb{N} \times G, +_I)$ is free of units.

Lemma 11.28. *The monoid $(\mathbb{N} \times G, +_I)$ is free of units if and only if $I(g, -g) \neq 0$ for all $g \in G \setminus \{0\}$.*

Proof. If g is an element in $G \setminus \{0\}$ such that $I(g, -g) = 0$, then $(0, g) +_I (0, -g) = (0, 0)$. This implies that $(0, g)$ is a nonzero unit and thus $(\mathbb{N} \times G, +_I)$ is not free of units.

If $(a, g), (b, h) \in \mathbb{N} \times G$ are such that $(a, g) +_I (b, h) = (0, 0)$, then $a + b + I(g, h) = 0$ and $g + h = 0$. As a, b and $I(g, h)$ are nonnegative integers, this implies that $a = b = I(g, h) = 0$. Note that $h = -g$. By hypothesis this implies that $g = 0$ and consequently $(a, g) = (0, 0)$. ☐

The quasi-Archimedean property is ensured when G is finite and $(\mathbb{N} \times G, +_I)$ is free of units.

Lemma 11.29. *If G is finite and $(\mathbb{N} \times G, +_I)$ is free of units, then $(\mathbb{N} \times G, +_I)$ is quasi-Archimedean.*

Proof. Recall that $(1,0)$ is an Archimedean element of $(\mathbb{N} \times G, +_I)$ (see the proof of Theorem 11.25). Note also that if a, b and c are elements in a monoid such that $a = b+c$, and b is Archimedean, then so is a.

For every $(a,g) \in \mathbb{N} \times G \setminus \{(0,0)\}$, $(a,g) = a(1,0) +_I (0,g)$. Hence if $a \neq 0$, then (a,g) is Archimedean. Thus assume that $a = 0$ (and consequently $g \neq 0$). As G is finite, there exist a positive integer such that $kg = 0$. Hence $-g = (k-1)g$. Note that $k(0,g) = (0,g) +_I (k-1)(0,g) = (0,g) +_I (b,-g) = (b+I(g,-g),0)$ for some nonnegative integer b. By Lemma 11.27, we know that $I(g,-g) \neq 0$ and thus $k(0,g) = (1,0) +_I (c,0)$ for some nonnegative integer c. This implies that $k(0,g)$ is Archimedean. Therefore $(0,g)$ is Archimedean. □

Let G be a finite group. The *order* of an element $g \in G$ is defined by $\mathrm{o}(g) = \min \{k \in \mathbb{N} \setminus \{0\} \mid kg = 0\}$. Recall that if $kg = 0$ for some positive integer k, then $k \equiv 0 \bmod \mathrm{o}(g)$ (just divide k by $\mathrm{o}(g)$ and let r be the remainder of the division; then $rg = 0$ and the minimality of $\mathrm{o}(g)$ forces r to be zero).

Lemma 11.30. *Let G be a finite group. Then $(\mathbb{Z} \times G, +_I)$ is torsion free if and only if $\sum_{i=1}^{\mathrm{o}(g)-1} I(g, ig) \not\equiv 0 \bmod \mathrm{o}(g)$ for all $g \in G \setminus \{0\}$.*

Proof. Let $g \in G \setminus \{0\}$. If $\sum_{i=1}^{\mathrm{o}(g)-1} I(g, ig) = \mathrm{o}(g)a$ for some $a \in \mathbb{N}$, then

$$\mathrm{o}(g)(-a,g) = \left(-\mathrm{o}(g)a + \sum_{i=1}^{\mathrm{o}(g)-1} I(g, ig), \mathrm{o}(g)g\right) = (0,0),$$

which implies that $(\mathbb{Z} \times G, +_I)$ is not torsion free.

Assume that $(\mathbb{Z} \times G, +_I)$ is not torsion free. Then there exist $(z,g) \in \mathbb{Z} \times G \setminus \{(0,0)\}$ and a positive integer k such that $k(z,g) = (0,0)$. This implies that $g \neq 0$, since $k(z,0) = (kz,0)$ which is nonzero for $z \neq 0$. Hence

$$\left(kz + \sum_{i=1}^{\mathrm{o}(g)-1} I(g, ig), kg\right) = (0,0).$$

This leads to $kg = 0$ and thus there exist $l \in \mathbb{N} \setminus \{0\}$ such that $k = \mathrm{o}(g)l$. Rewriting the first coordinate of the above equality, we get

$$\mathrm{o}(g)lz + \sum_{i=1}^{\mathrm{o}(g)l-1} I(g, ig) - 0.$$

So

$$\mathrm{o}(g)l = l \sum_{i=1}^{\mathrm{o}(g)-1} I(g, ig) = 0$$

and consequently

$$\sum_{i=1}^{\mathrm{o}(g)-1} I(g, ig) \equiv 0 \bmod \mathrm{o}(g). \qquad \square$$

Note that every submonoid of a torsion free group is torsion free. Thus by Theorem 11.18, with this last lemma we already have all the ingredients to describe the conditions we should impose on I and G so that $(\mathbb{N} \times G, +_I)$ is isomorphic to a numerical semigroup.

Theorem 11.31. *Let G be a finite cyclic group and let $I : G \times G \to \mathbb{N}$ be a map fulfilling*

P.1) $I(g_1, g_2) = I(g_2, g_1)$ for all $g_1, g_2 \in G$,
P.2) $I(g_1, g_2) + I(g_1 + g_2, g_3) = I(g_2, g_3) + I(g_1, g_2 + g_3)$ for all $g_1, g_2, g_3 \in G$,
P.3) $I(0,0) = 0$,
P.4) $I(g, -g) \neq 0$ for all $g \in G \setminus \{0\}$,
P.5) $\sum_{i=1}^{o(g)-1} I(g, ig) \not\equiv 0 \bmod o(g)$ for all $g \in G \setminus \{0\}$.

Then $(\mathbb{N} \times G, +_I)$ is isomorphic to a numerical semigroup. Moreover, every numerical semigroup is isomorphic to a monoid of this form.

Proof. We know by Theorem 11.25 that $(\mathbb{N} \times G, +_I)$ is an $\mathfrak{N}$-monoid. Hence it is cancellative. Lemma 11.27 tells us that it is finitely generated. As a consequence of Lemmas 11.28 and 11.29, this monoid is quasi-Archimedean. Lemma 11.30 asserts that it is torsion free. Hence by Theorem 11.18 it is isomorphic to a numerical semigroup.

Every numerical semigroup S is an $\mathfrak{N}$-monoid and by Theorem 11.25 there exist a group G and I fulfilling P.1), P.2) and P.3) such that S is isomorphic to $(\mathbb{N} \times G, +_I)$. By Proposition 11.26, G is cyclic with finitely many elements. Lemmas 11.28 and 11.29 ensure that Conditions P.4) and P.5) hold.   □

Exercises

Exercise 11.1 ([80]). Let S be a finitely generated monoid. Prove that S is cancellative and torsion free if and only if it is isomorphic to a submonoid of $(\mathbb{Z}^k, +)$ for some positive integer k. (*Hint:* As S is finitely generated, S is isomorphic to $\mathbb{N}^n / \sigma$ for some congruence σ over $\mathbb{N}$. The set $\{a - b \mid (a,b) \in \sigma\}$ is a subgroup of $\mathbb{Z}^n$, and the columns of its defining equations generate a submonoid isomorphic to S.)

Exercise 11.2 ([80]). Let S be a finitely generated monoid. Prove that S is cancellative, torsion free and free of units if and only if it is isomorphic to a submonoid of $(\mathbb{N}^k, +)$ for some positive integer k.

Exercise 11.3 ([80]). Let S be the monoid

$$\{(x,y) \in \mathbb{Z}^2 \mid x > 0\} \cup \{(0,y) \in \mathbb{Z}^2 \mid y \geq 0\}.$$

Prove that S is cancellative, torsion free and free of units. Show that there is no $k \in \mathbb{N}$ such that S is isomorphic to a submonoid of $\mathbb{N}^k$.

Exercise 11.4. Let $S = ((1,1) + \mathbb{N}^2) \cup \{(0,0)\}$. Prove that S is quasi-Archimedean, torsion free, cancellative, free of units, with only one idempotent, and it is not isomorphic to any numerical semigroup.

Exercise 11.5. Let S be a finitely generated submonoid of $\mathbb{N}^k$ for some positive integer k. Prove that S is isomorphic to a numerical semigroup if and only if it is quasi-Archimedean (that is, numerical semigroups are the only quasi-Archimedean affine semigroups).

Exercise 11.6. Let S be a nontrivial submonoid of $\mathbb{N}^k$ for some positive integer k. Prove that S is isomorphic to a numerical semigroup if and only if it is hereditarily finitely generated.

Exercise 11.7. Find a finitely generated monoid and a submonoid of this monoid that it is not finitely generated.

Exercise 11.8 ([75]). Let $X = \{x_1, \ldots, x_n\}$ and let σ be a congruence over $\mathrm{Free}(X)$. Prove that $\frac{\mathrm{Free}(X)}{\sigma}$ is hereditarily finitely generated if and only if for all $i, j \in \{1, \ldots, n\}$ with $i \neq j$, there exists an element in $\sigma \setminus \Delta(\mathrm{Free}(X))$ of the form

$$(a_{i_j} x_i + (b_{i_j} + 1)x_j, c_{i_j} x_i + x_j).$$

Exercise 11.9. Find a cancellative, free of units, hereditarily finitely generated monoid that is not torsion free.

Exercise 11.10. Let X be a set and let $\mathscr{P}(X)$ be the set of subsets of X. Prove that $(\mathscr{P}(X), \cap)$ is a torsion free monoid that in general is not cancellative.

Exercise 11.11 ([80]). Prove that every finitely generated monoid is the union of finitely many Archimedean semigroups. (*Hint:* For $x, y \in S$ define $x \mathfrak{N} y$ if there exist $z \in S$ and $n \in \mathbb{N}$ such that $nx = y + z$. This is an equivalence relation and its equivalence classes are the Archimedean semigroups we are looking for.)

Bibliography

1. S. S. Abhyankar, Local rings of high embedding dimension, Amer. J. Math. **89** (1967), 1073–1077.
2. R. Apéry, Sur les branches superlinéaires des courbes algébriques, C. R. Acad. Sci. Paris **222** (1946), 1198–1200.
3. C. Arf, Une interprétation algébrique de la suite des ordres de multiplicité d'une branche algébrique, Proc. London Math. Soc. (2) **50** (1948), 256–287.
4. V. Barucci, Numerical semigroup algebras, in Multiplicative Ideal Theory in Commutative Algebra, pp. 39–53, Springer, New York, 2006, Editors: James W. Brewer, Sarah Glaz, William J. Heinzer and Bruce M. Olberding.
5. V. Barucci, D. E. Dobbs, M. Fontana, Maximality properties in numerical semigroups and applications to one-dimensional analytically irreducible local domains, Memoirs of the Amer. Math. Soc. **598** (1997).
6. J. Bertin, P. Carbonne, Semi groupes d'entiers et application aux branches, J. Algebra **49** (1977), 81–95.
7. M. Bras-Amorós, Improvements to evaluation codes and new characterizations of Arf semigroups, in Applied Algebra, Algebraic Algorithms and Error-Correcting Codes (Toulouse, 2003), pp. 204–215, Lecture Notes in Comput. Sci., 2643, Springer, Berlin, 2003, Editors: M. Fossorier, T. Høholdt and A. Poli.
8. M. Bras-Amorós, Acute semigroups, the order bound on the minimum distance, and the Feng-Rao improvements, IEEE Trans. Inform. Theory **50** (2004), no. 6, 1282–1289.
9. M. Bras-Amorós, A note on numerical semigroups, IEEE Trans. Inform. Theory **53** (2007), no. 2, 821–823.
10. M. Bras-Amorós, Fibonacci-like behavior of the number of numerical semigroups of a given genus, Semigroup Forum **76** (2008), 379–384.
11. M. Bras-Amorós, P. A. García-Sánchez, Patterns on numerical semigroups, Linear Algebra Appl. **414** (2006), 652–669.
12. A. Brauer, On a problem of partitions, Amer. J. Math. **64** (1942), 299–312.
13. A. Brauer, J. E. Shockley, On a problem of Frobenius, J. Reine Angew. Math. **211** (1962), 215–220.
14. H. Bresinsky, On prime ideals with generic zero $x_i = t^{n_i}$, Proc. Amer. Math. Soc. **47** (1975), 329–332.
15. W. C. Brown, F. Curtis, Numerical semigroups of maximal and almost maximal length, Semigroup Forum **42** (1991), 219–235.
16. W. C. Brown, J. Herzog, One-dimensional local rings of maximal and almost maximal length, J. Algebra **151** (1992), 332–347.
17. A. Campillo, On saturation of curve singularities (any characteristic), Proc. of Symp. in Pure Math. **40** (1983), 211–220.

J.C. Rosales, P.A. García-Sánchez, *Numerical Semigroups*,
Developments in Mathematics 20, DOI 10.1007/978-1-4419-0160-6,
© Springer Science+Business Media, LLC 2009

18. A. Campillo, J. I. Farrán, C. Munuera, On the parameters of algebraic-geometry codes related to Arf semigroups, IEEE Trans. Inform. Theory **46** (2000), no. 7, 2634–2638.

19. J. Castellanos, A relation between the sequence of multiplicities and the semigroups of values of an algebroid curve, J. Pure Appl. Algebra **43** (1986), 119–127.

20. S. T. Chapman, P. A. García-Sánchez, D. Llena, V. Ponomarenko, J. C. Rosales, The catenary and tame degree in finitely generated cancellative monoids, Manuscripta Math. **120** (2006), 253–264.

21. F. Curtis, On formulas for the Frobenius number of a numerical semigroup, Math. Scand. **67** (1990), no. 2, 190–192.

22. J. L. Davison, On the linear Diophantine problem of Frobenius, J. Number Theory **48** (1994), 353–363.

23. M. Delgado, P. A. García-Sánchez, J. Morais, "numericalsgps": a GAP package on numerical semigroups. (http://www.gap-system.org/Packages/numericalsgps.html).

24. M. Delgado, P. A. García-Sánchez, J. C. Rosales, J. M. Urbano-Blanco, Systems of proportionally modular Diophantine inequalities, Semigroup Forum **76** (2008), 469–488.

25. M. Delgado, J. C. Rosales, On the Frobenius number of a proportionally modular Diophantine inequality, Port. Math. (N.S.) **63** (2006), no. 4, 415–425.

26. F. Delgado, A. Núñez, Monomial rings and saturated rings, in Géométrie Algébrique et Applications, I (La Rábida, 1984), pp. 23–34, Travaux en Cours, 22, Hermann, Paris, 1987.

27. C. Delorme, Sous-monoïdes d'intersection complète de $\mathbb{N}$, Ann. Scient. École Norm. Sup. (4) **9** (1976), 145–154.

28. M. Djawadi, G. Hofmeister, Linear Diophantine problems, Arch. Math. (Basel) **66** (1996), 19–29.

29. L. Fel, Frobenius problem for semigroups $S(d_1, d_2, d_3)$, Functional Analysis and Other Mathematics **1** (2006), 119–157.

30. K. G. Fischer, W. Morris, J. Shapiro, Affine semigroup rings that are complete intersections, Proc. Amer. Math. Soc. **125** (1997), 3137–3145.

31. P. Freyd, Rédei's finiteness theorem for commutative semigroups, Proc. Amer. Math. Soc. **19** (1968), 1003.

32. R. Fröberg, C. Gottlieb, R. Häggkvist, On numerical semigroups, Semigroup Forum **35** (1987), 63–83.

33. P. A. García-Sánchez, J. C. Rosales, Numerical semigroups generated by intervals, Pacific J. Math. **191** (1999), no. 1, 75–83.

34. P. A. García-Sánchez, J. C. Rosales, Every positive integer is the Frobenius number of an irreducible numerical semigroup with at most four generators, Arkiv Mat. **42** (2004), 301–306.

35. The GAP Group. GAP - Groups, Algorithms, and Programming, Version 4.4, 2004. (http://www.gap-system.org).

36. A. Geroldinger, F. Halter-Koch, Non-Unique Factorizations: Algebraic, Combinatorial and Analytic Theory, Pure and Applied Mathematics, 278, Chapman & Hall/CRC, Boca Raton, 2006.

37. R. Gilmer, Commutative Semigroup Rings, Chicago Lectures in Mathematics, University of Chicago Press, Chicago, 1984.

38. R. L. Graham, D. E. Knuth, O. Patashnik, Concrete Mathematics: A Foundation for Computer Science, Addison-Wesley Publishing Company, Advanced Book Program, Reading, MA, 1989.

39. P. A. Grillet, A short proof of Rédei's theorem, Semigroup Forum **46** (1993), 126–127.

40. G. H. Hardy, E. M. Wright, An Introduction to the Theory of Numbers, Oxford University Press, Oxford, 1979.

41. J. Herzog, Generators and relations of abelian semigroups and semigroup rings, Manuscripta Math. **3** (1970), 175–193.

42. S. M. Johnson, A linear Diophantine problem, Canad. J. Math. **12** (1960), 390–398.

43. C. Kirfel, R. Pellikaan, The minimum distance of codes in an array coming from telescopic semigroups, IEEE Trans. Inform. Theory **41** (1995), no. 6, part 1, 1720–1732. Special issue on algebraic geometry codes.

44. E. Kunz, The value-semigroup of a one-dimensional Gorenstein ring, Proc. Amer. Math. Soc. **25** (1970), 748–751.

45. E. Kunz, Über die Klassifikation numerischer Halbgruppen, Regensburger Mathematische Schriften, 11, Universität Regensburg, Fachbereich Mathematik, Regensburg, 1987.

46. R. G. Levin, On commutative, nonpotent Archimedean semigroups, Pacific J. Math. **27** (1968), 365–371.

47. J. Lipman, Stable ideals and Arf rings, Amer. J. Math. **93** (1971), 649–685.

48. D. Narsingh, Graph Theory with Applications to Engineering and Computer Science, Prentice Hall Series in Automatic Computation, Prentice Hall, Englewood Cliffs, NJ, 1974.

49. A. Núñez, Algebro-geometric properties of saturated rings, J. Pure Appl. Algebra **59** (1989), 201–214.

50. F. Pham, B. Teissier, Fractions lipschitziennes et saturation de Zariski des algèbres analytiques complexes, Centre Math. École Polytech., Paris, 1969, in Actes du Congrès International des Mathématiciens (Nice, 1970), Tome 2, pp. 649–654, Gauthier-Villars, Paris, 1971.

51. G. B. Preston, Rédei's characterization of congruences on finitely generated free commutative semigroups, Acta Math. Acad. Sci. Hungar. **26** (1975), 337–342.

52. J. L. Ramírez Alfonsín, The Diophantine Frobenius Problem, Oxford University Press, Oxford, 2005.

53. L. Rédei, The Theory of Finitely Generated Commutative Semigroups, Pergamon, Oxford-Edinburgh-New York, 1965.

54. A. M. Robles-Pérez, J. C. Rosales, Equivalent proportionally modular Diophantine inequalities, Archiv der Mathematik **90** (2008), 24–30.

55. J. C. Rosales, Semigrupos numéricos, Tesis Doctoral, Universidad de Granada, Spain, 2001.

56. J. C. Rosales, Function minimum associated to a congruence on integral n-tuple space, Semigroup Forum **51** (1995) 87–95.

57. J. C. Rosales, On numerical semigroups, Semigroup Forum **52** (1996), 307–318.

58. J. C. Rosales, On symmetric numerical semigroups, J. Algebra **182** (1996), no. 2, 422–434.

59. J. C. Rosales, An algorithmic method to compute a minimal relation for any numerical semigroup, Internat. J. Algebra Comput. **6** (1996), no. 4, 441–455.

60. J. C. Rosales, On presentations of subsemigroups of $\mathbb{N}^n$, Semigroup Forum **55** (1997), 152–159.

61. J. C. Rosales, Numerical semigroups with Apéry sets of unique expression, J. Algebra **226** (2000), no. 1, 479–487.

62. J. C. Rosales, Symmetric numerical semigroups with arbitrary multiplicity and embedding dimension, Proc. Amer. Math. Soc. **129** (2001), no. 8, 2197–2203.

63. J. C. Rosales, Principal ideals of numerical semigroups, Bull. Belg. Math. Soc. Simon Stevin **10** (2003), 329–343.

64. J. C. Rosales, Fundamental gaps of numerical semigroups generated by two elements, Linear Algebra Appl. **405** (2005), 200–208.

65. J. C. Rosales, Adding or removing an element from a pseudo-symmetric numerical semigroup, Boll. Unione Mat. Ital. **9-B** (2006), 681–696.

66. J. C. Rosales, Subadditive periodic functions and numerical semigroups, J. Algebra Appl. **6** (2007), no. 2, 305–313.

67. J. C. Rosales, Families of numerical semigroups closed under finite intersections and for the Frobenius number, Houston J. Math. **34** (2008), 339–348.

68. J. C. Rosales, Numerical semigroups that differ from a symmetric numerical semigroup in one element, Algebra Colloquium **15** (2008), 23–32.

69. J. C. Rosales, One half of a pseudo-symmetric numerical semigroup, Bull. London Math. Soc. **40** (2008), 347–352.

70. J. C. Rosales, M. B. Branco, Decomposition of a numerical semigroup as an intersection of irreducible numerical semigroups, Bull. Belg. Math. Soc. Simon Stevin **9** (2002), 373–381.

71. J. C. Rosales, M. B. Branco, Numerical semigroups that can be expressed as an intersection of symmetric numerical semigroups, J. Pure Appl. Algebra **171** (2002), nos. 2–3, 303–314.

72. J. C. Rosales, M. B. Branco, Three families of free numerical semigroups with arbitrary embedding dimension, Int. J. Commutative Rings **1** (2002), no. 4, 195–201. Also in the book: Ayman Badawi, Commutative Rings, pp. 151–157, Nova Science Publishers, New York, 2002.

73. J. C. Rosales, M. B. Branco, Irreducible numerical semigroups, Pacific J. Math. **209** (2003), 131–143.

74. J. C. Rosales, M. B. Branco, Irreducible numerical semigroups with arbitrary multiplicity and embedding dimension, J. Algebra **264** (2003), 305–315.

75. J. C. Rosales, J. I. García-García, Hereditarily finitely generated commutative monoids, J. Algebra **221** (1999), no. 2, 723–732.

76. J. C. Rosales, J. I. García-García, Generalized N-semigroups, in Proceedings of the International Conference on Semigroups, pp. 171–180, World Scientific, Singapore, 2000, Editors: Paula Smith, Emília Giraldes and Paula Martins.

77. J. C. Rosales, P. A. García-Sánchez, On complete intersection affine semigroups, Comm. Algebra **23** (1995), 5395–5412.

78. J. C. Rosales, P. A. García-Sánchez, On Cohen-Macaulay and Gorenstein simplicial affine semigroups, Proc. Edinburgh Math. Soc. **41** (1998), 517–537.

79. J. C. Rosales, P. A. García-Sánchez, On numerical semigroups with high embedding dimension, J. Algebra **203** (1998), no. 2, 567–578.

80. J. C. Rosales, P. A. García-Sánchez, Finitely Generated Commutative Monoids, Nova Science Publishers, New York, 1999.

81. J. C. Rosales, P. A. García-Sánchez, Numerical semigroups with embedding dimension three, Archiv der Mathematik **83** (2004), 488–496.

82. J. C. Rosales, P. A. García-Sánchez, Pseudo-symmetric numerical semigroups with three generators, J. Algebra **291** (2005), 46–54.

83. J. C. Rosales, P. A. García-Sánchez, Every numerical semigroup is one half of infinitely many symmetric numerical semigroups, Comm. Algebra **36** (2008), 2910–2916.

84. J. C. Rosales, P. A. García-Sánchez, Numerical semigroups having a Toms decomposition, Canadian Math. Bull. **51** (2008), 134–139.

85. J. C. Rosales, P. A. García-Sánchez, J. I. García-García, Every positive integer is the Frobenius number of a numerical semigroup with three generators, Math. Scand. **94** (2004), 5–12.

86. J. C. Rosales, P. A. García-Sánchez, J. I. García-García, k-factorized elements in telescopic numerical semigroups, in Proceedings of the Chapel Hill Conference on Arithmetical Properties of Commutative Rings and Monoids, pp. 260–271, Lecture Notes in Pure Appl. Math., 241, CRC Press, Boca Raton, 2005, Editor: S. T. Chapman.

87. J. C. Rosales, P. A. García-Sánchez, J. I. García-García, M. B. Branco, Systems of inequalities and numerical semigroups, J. London Math. Soc. **65** (2002), no. 3, 611–623.

88. J. C. Rosales, P. A. García-Sánchez, J. I. García-García, M. B. Branco, Arf numerical semigroups, J. Algebra **276** (2004), 3–12.

89. J. C. Rosales, P. A. García-Sánchez, J. I. García-García, M. B. Branco, Saturated numerical semigroups, Houston J. Math. **30** (2004), 321–330.

90. J. C. Rosales, P. A. García-Sánchez, J. I. García-García, J. A. Jimenez-Madrid, The oversemigroups of a numerical semigroup, Semigroup Forum **67** (2003), 145–158.

91. J. C. Rosales, P. A. García-Sánchez, J. I. García-García, J. A. Jiménez-Madrid, Fundamental gaps in numerical semigroups, J. Pure Appl. Alg. **189** (2004), 301–313.

92. J. C. Rosales, P. A. García-Sánchez, J. I. García-García, J. M. Urbano-Blanco, Proportionally modular Diophantine inequalities, J. Number Theory **103** (2003), 281–294.

93. J. C. Rosales, P. A. García-Sánchez, J. M. Urbano-Blanco, On presentations of commutative monoids, Internat. J. Algebra Comput. **9** (1999), no. 5, 539–553.

94. J. C. Rosales, P. A. García-Sánchez, J. M. Urbano-Blanco, Modular Diophantine inequalities and numerical semigroups, Pacific J. Math. **218** (2005), no. 2, 379–398.

95. J. C. Rosales, P. A. García-Sánchez, J. M. Urbano-Blanco, The set of solutions of a proportionally modular Diophantine inequality, J. Number Theory. **128** (2008), 453–467.

96. J. C. Rosales, P. Vasco, The smallest positive integer that is solution of a proportionally modular Diophantine inequality, Math. Inequal. Appl. **11** (2008), 203–212.

97. J. C. Rosales, J. M. Urbano-Blanco, Opened modular numerical semigroups, J. Algebra **306** (2006), 368–377.

98. J. C. Rosales, J. M. Urbano-Blanco, Proportionally modular Diophantine inequalities and full semigroups, Semigroup Forum **72** (2006), 362–374.

99. J. D. Sally, On the associated graded ring of a local Cohen-Macaulay ring, J. Math. Kyoto Univ. **17** (1977), 19–21.

100. J. D. Sally, Cohen-Macaualy local rings of maximal embedding dimension, J. Algebra **56** (1979), 168–183.

101. E. S. Selmer, On a linear Diophantine problem of Frobenius, J. Reine Angew. Math. **293/294** (1977), 1–17.

102. J. J. Sylvester, Mathematical questions with their solutions, Educational Times **41** (1884), 21.

103. T. Tamura, Nonpotent Archimedean semigroups with cancellative law, I, J. Gakugei Tokushima Univ. **8** (1957), 5–11.

104. J. M. Urbano-Blanco, Semigrupos numéricos proporcionalmente modulares, Tesis Doctoral, Universidad de Granada, Spain, March 2005.

105. B. Teissier, Appendice à "Le probléme des modules pour le branches planes", cours donné par O. Zariski au Centre de Math. de L'École Polytechnique, Paris (1973).

106. A. Toms, Strongly perforated K_0-groups of simple C^*-algebras, Canad. Math. Bull. **46** (2003), 457–472.

107. K. Watanabe, Some examples of one dimensional Gorenstein domains, Nagoya Math. J. **49** (1973), 101–109.

108. H. S. Wilf, A circle-of-lights algorithm for the "money-changing problem", Amer. Math. Monthly **85** (1978), 562–565.

109. O. Zariski, General theory of saturation and saturated local rings I, II, III, Amer. J. Math. **93** (1971), 573–684, 872–964, **97** (1975), 415–502.

List of symbols

$\langle n_1, \ldots, n_p \rangle$ - submonoid generated by $\{n_1, \ldots, n_p\}$, p. 1.
$\mathrm{Arf}(S)$ - Arf closure of the numerical semigroup S, p. 25.
$\mathrm{Ap}(S, m)$ - Apéry set of the element $m \neq 0$ in S, p. 8.
$\mathrm{Ch}(S)$ - chain associated to the numerical semigroup S, p. 92.
$\mathrm{Ch}(X)$ - union of the chains associated to the numerical semigroups in X, p. 98.
$\mathrm{Cong}(\rho)$ - congruence generated by ρ, p. 108.
$\mathrm{d}_A(a)$ - greatest common divisor of the elements in A less than or equal
to $a \in A$, p. 28.
$\mathrm{D}(X)$ - set of all positive divisors of the elements of X, p. 51.
$\Delta(X) = \{ (x,x) \mid x \in X \}$ - the diagonal of $X \times X$, p. 108.
$\mathrm{e}(S)$ - embedding dimension of the numerical semigroup S, p. 9.
$\mathrm{F}(S)$ - Frobenius number of the numerical semigroup S, p. 9.
$\mathrm{FG}(S)$ - set of fundamental gaps of S, p. 52.
$\mathrm{Free}(X)$ - free monoid on X, p. 107.
$\mathscr{F}(X)$ - Frobenius variety generated by the set of numerical semigroups X, p. 98.
$\mathrm{g}(S)$ - genus (or degree of singularity) of the numerical semigroup S, p. 9.
$\mathscr{G}(\mathscr{V})$ - graph associated to the Frobenius variety $\mathscr{V}$, pp. 92, 101.
$\mathscr{I}(S)$ - set of irreducible numerical semigroups containing the numerical semigroup
S, p. 47.
$\mathrm{im}(f)$ - image of the homomorphism f, p. 106.
$\mathrm{Irr}(\sigma)$ - set of irreducible elements of σ, p. 109.
$\mathrm{J}(S)$ - for a numerical semigroup minimally generated by $n_1 < \cdots < n_e$ is the set
$\{ \lambda_2 x_2 + \cdots + \lambda_e x_e \mid \lambda_2 n_2 + \cdots + \lambda_e n_e \notin \mathrm{Ap}(S, n_1) \}$, p. 138.
$\mathrm{ker}(f)$ the kernel congruence of the monoid homomorphism f, p. 106.
$\mathrm{m}(S)$ - multiplicity of the numerical semigroup S, p. 9.
$\mathrm{M}(f)$ - monoid associated to the subadditive function f, p. 58.
$\mathrm{Maximals}_\leq(X)$ - maximal elements of X with respect to the ordering $\leq$, p. 13.
$\mathrm{Minimals}_\leq(X)$ - minimal elements of X with respect to the ordering $\leq$, p. 13.
$a \bmod b$ - is the quotient of the division of a by b;
$a \equiv b \bmod m$ means $(a - b) \bmod m = 0$, p. 20.
$\mathrm{n}(S)$ - cardinality of the set of elements in S less than its Frobenius number, p. 15.

$N(S)$ - set of elements in S less than its Frobenius number, p. 15.

$\mathbb{N}$ - set of nonnegative integers, p. 1.

$o(g)$ - order of an element g in a group G, p. 167.

$\mathscr{O}(S)$ - set of numerical semigroups containing the numerical semigroup S, p. 44.

$PF(S)$ - set of pseudo-Frobenius numbers of the numerical semigroup S, p. 13.

$\mathbb{Q}_0^+$ - set of nonnegative rational numbers, p. 59.

$Q(S)$ - quotient group of S, p. 163.

$\mathscr{S}$ - set of all numerical semigroups, p. 47.

$\mathscr{S}_m$ - set of numerical semigroups with multiplicity m, p. 58.

$\mathscr{S}(g_1, \ldots, g_t)$ - set of numerical semigroups not cutting $\{g_1, \ldots, g_t\}$, p. 47.

$\mathscr{S}(P)$ - set of numerical semigroups admitting a pattern P, p. 96.

$S(a, b, c)$ - set of integer solutions to $ax \bmod b \leq cx$, p. 58.

$S(A)$ - with $A \subset \mathbb{Q}_0^+$, the set of integers of the submonoid $\langle A \rangle$, p. 59.

$Sat(S)$ - saturated closure of the numerical semigroup S, p. 30.

$\mathscr{S}\mathscr{F}_m$ - set of m-periodic subadditive functions, p. 58.

$SG(S)$ - set of special gaps of S, p. 44.

R - relation defining the R-classes of the expressions of a given element in a numerical semigroup, p. 111.

$t(S)$ - type of the numerical semigroup S, p. 13.

$\mathscr{V}$ - a Frobenius variety, p. 99.

$\mathscr{V}(A)$ - $\mathscr{V}$-monoid generated by A, p. 99. Used also as a prefix to denote systems of generators and monoids relative to this variety; see Chapter 6.

$\mathbb{Z}$ - set of integers, p. 13.

$Z(n)$ - set of factorizations or expression of n in a numerical semigroup S, p. 111.

$Z_B(n)$ - set of factorizations or expression of n in a numerical semigroup S relative to the set of generators B, p. 111.

Index

Symbols

0-matrix   150

A

adjacent fractions   66
Apéry set   8
   with respect to a set   124
Archimedean element   155
Arf closure
   of a semigroup   25
   of a set   25

B

Bézout sequence   61
   end of a   61
   lenght of a   61
   proper   65
binary relation   106
   equivalence   106
   inverse relation   108

C

chain associated to a numerical semigroup
   92
complete intersection   129
conductor   9
congruence
   cancellative   109
   diagonal congruence   108
   finitely generated   108
   generated by a set   108
   minimal relation   110
   monoid   106

semigroup   157

D

degree of singularity   9
Dickson's Lemma   109
directed graph   91
   of all numerical semigroups   92
   edges   91
   path   91
   vertex   91
dominant   17

E

embedding dimension   9

F

factorization   122
   catenary degree   122
   distance   122
   greatest common divisor   122
   length   122
Frobenius number   9
Frobenius variety   93
   generated by a family of numerical
      semigroups   98

G

gap   9
   fundamental   52
   special   44
genus   9
gluing   124
   of numerical semigroups   130

graph   111
    associated to an element in a numerical
        semigroup   113
    connected   111
        generating tree   111
    connected component   113
    directed   91
    edge   111
    path   111
    vertices   111

I

ideal   17
    canonical   18
    maximal   17
    principal   23
    relative   17
idempotent element   155
irreducible element of a congruence   108

M

minimal presentation   111
minimal system of generators   8
modular Diophantine inequality   69
monoid   6
    cancellative   109
    epimorphism   6
    finitely generated   6
    finitely presented   108
    free   107
    free of units   159
    half-factorial   122
    hereditarily finitely generated   161
    homomorphism   6
        image   106
        kernel congruence   106
    isomorphism   6
    monomorphism   6
    𝔑-monoid   163
    quasi-Archimedean   159
    quotient   106
    weakly cancellative   161
multiplicity   9

N

numerical semigroup   6
    acute   17
    admitting a pattern   96
    almost symmetric   55
    Arf   23
    arithmetic   88

associated to a set of rational numbers   59
    free   133
    half-line   17
    irreducible   33
    maximal embedding dimension   20
    modular   69
    ordinary   17
    proportionally modular   59
    pseudo-symmetric   34
    saturated   28
    simple   154
    symmetric   34
    telescopic   136
    with Apéry set of unique expression   138

O

order   167
oversemigroup   44

P

pattern   96
    admissible   97
    strongly admissible   97
presentation   108
proportionally modular Diophantine inequality
        59
    factor of a   59
    modulus of a   59
    proportion   59
pseudo-Frobenius number   13

Q

quotient group   163
quotient of a numerical semigroup by an
        integer   78
quotient set   106

R

ratio   17
R-class   111

S

Saturated closure   30
semigroup   5
    Archimedean   155
    epimorphism   6
    finitely generated   6
    homomorphism   6
    isomorphism   6
    monomorphism   6
    multiple joined   155
    quotient   157

system of generators of a   6
  torsion free   159
set of expressions   111
strongly positive sequence   150
subadditive function   57
  period of a   58
submonoid   6
  trivial   6
subsemigroup   5
  subsemigroup generated by a set   6
system of generators
  of a congruence   108
  of a monoid   6

T

tree   91
  leaf   92

root   91
son   92
undirected   111
type   13

U

unique expression element   138
unit   159

V

$\mathcal{V}$-monoid   99

W

weight
  modular Diophantine inequality   75